Janin Hofmann

Requirements of T cell-driven Autoimmunity in Alymphoplastic Mice

Janin Hofmann

Requirements of T cell-driven Autoimmunity in Alymphoplastic Mice

Dissertation, University Zurich, Experimental Immunology, 2011

Südwestdeutscher Verlag für Hochschulschriften

Impressum/Imprint (nur für Deutschland/only for Germany)
Bibliografische Information der Deutschen Nationalbibliothek: Die Deutsche Nationalbibliothek verzeichnet diese Publikation in der Deutschen Nationalbibliografie; detaillierte bibliografische Daten sind im Internet über http://dnb.d-nb.de abrufbar.
Alle in diesem Buch genannten Marken und Produktnamen unterliegen warenzeichen-, marken- oder patentrechtlichem Schutz bzw. sind Warenzeichen oder eingetragene Warenzeichen der jeweiligen Inhaber. Die Wiedergabe von Marken, Produktnamen, Gebrauchsnamen, Handelsnamen, Warenbezeichnungen u.s.w. in diesem Werk berechtigt auch ohne besondere Kennzeichnung nicht zu der Annahme, dass solche Namen im Sinne der Warenzeichen- und Markenschutzgesetzgebung als frei zu betrachten wären und daher von jedermann benutzt werden dürften.

Coverbild: www.ingimage.com

Verlag: Südwestdeutscher Verlag für Hochschulschriften GmbH & Co. KG
Heinrich-Böcking-Str. 6-8, 66121 Saarbrücken, Deutschland
Telefon +49 681 37 20 271-1, Telefax +49 681 37 20 271-0
Email: info@svh-verlag.de

Approved by: Schweiz, Zuerich, Universitaet Zuerich, Dissertation, 2011

Herstellung in Deutschland:
Schaltungsdienst Lange o.H.G., Berlin
Books on Demand GmbH, Norderstedt
Reha GmbH, Saarbrücken
Amazon Distribution GmbH, Leipzig
ISBN: 978-3-8381-3208-2

Imprint (only for USA, GB)
Bibliographic information published by the Deutsche Nationalbibliothek: The Deutsche Nationalbibliothek lists this publication in the Deutsche Nationalbibliografie; detailed bibliographic data are available in the Internet at http://dnb.d-nb.de.
Any brand names and product names mentioned in this book are subject to trademark, brand or patent protection and are trademarks or registered trademarks of their respective holders. The use of brand names, product names, common names, trade names, product descriptions etc. even without a particular marking in this works is in no way to be construed to mean that such names may be regarded as unrestricted in respect of trademark and brand protection legislation and could thus be used by anyone.

Cover image: www.ingimage.com

Publisher: Südwestdeutscher Verlag für Hochschulschriften GmbH & Co. KG
Heinrich-Böcking-Str. 6-8, 66121 Saarbrücken, Germany
Phone +49 681 37 20 271-1, Fax +49 681 37 20 271-0
Email: info@svh-verlag.de

Printed in the U.S.A.
Printed in the U.K. by (see last page)
ISBN: 978-3-8381-3208-2

Copyright © 2012 by the author and Südwestdeutscher Verlag für Hochschulschriften GmbH & Co. KG and licensors
All rights reserved. Saarbrücken 2012

DISCLAIMER

This thesis was based upon and partly adapted from the following publications:

Neo-lymphoid aggregates in the adult liver can initiate potent cell-mediated immunity.
Greter M, Hofmann J, Becher B.
PLoS Biol. 2009 May 5;7(5)

B cells need a proper house, whereas T cells are happy in a cave: the dependence of lymphocytes on secondary lymphoid tissues during evolution.
Hofmann J, Greter M, Du Pasquier L, Becher B.
Trends Immunol. 2010 Apr;31(4)

NIK signaling in dendritic cells but not in T cells is required for the development of effector T cells and cell-mediated immune responses.
Hofmann J, Mair F, Greter M, Schmidt-Supprian M and Becher B
J Exp Med. 2011 Aug 29;208(9)

TABLE OF CONTENTS

DISCLAIMER 3

TABLE OF CONTENTS 5

ABBREVIATIONS 9

ZUSAMMENFASSUNG 13

SUMMARY 15

INTRODUCTION 17
   THE IMMUNE SYSTEM 17
   LYMPHOID TISSUES: WHERE T CELLS DEVELOP AND POLARIZE 18
      PRIMARY LYMPHOID TISSUES 18
         THYMIC T CELL DEVELOPMENT 19
      SECONDARY LYMPHOID TISSUES 21
         T CELL ACTIVATION AND POLARIZATION 22
      TERTIARY LYMPHOID TISSUES 24
   SELF TOLERANCE AND AUTOIMMUNITY 25
      MULTIPLE SCLEROSIS 26
         PATHOLOGY AND SYMPTOMS 26
         EPIDEMIOLOGY 26
         RISK FACTORS 26
         THERAPIES 27
      EXPERIMENTAL AUTOIMMUNE ENCEPHALOMYELITIS (EAE) 29
   NFκB-SIGNALING 30
      CANONICAL NFκB-SIGNALING 31
      NON-CANONICAL NFκB-SIGNALING AND NIK 32
      NFκB AND NIK PLAY A ROLE IN VARIOUS DISEASES 33
   ALYMPHOPLASIA MICE 34

SPECIFIC AIMS 37

RESULTS 39
   NEO-LYMPHOID AGGREGATES IN THE ADULT LIVER CAN INITIATE POTENT CELL-MEDIATED IMMUNITY 39

AUTOIMMUNITY CANNOT BE INITIATED IN NIK$^{ALY/ALY}$ MICE 39
INDUCTION OF PRODUCTIVE T CELL IMMUNITY IN THE ABSENCE OF SLTS 41
B AND T CELLS HAVE DIFFERENT STRUCTURAL REQUIREMENTS FOR PRIMING AND MATURATION 43
IN THE ABSENCE OF SLTS, SUBCUTANEOUSLY DELIVERED AG IS TRANSPORTED INTO THE LIVER 45
EXTRA-LYMPHOID AGGREGATES IN THE LIVER HOST T CELL/APC ENCOUNTERS 48
THE ADULT LIVER CAN SUPPORT T CELL, BUT NOT B CELL PRIMING 52
PRIMING OF CYTOTOXIC ANTITUMOR T CELLS IS INDEPENDENT OF SLTS 53
LIVER FOLLICLES ARE INDUCED BY IMMUNIZATION AND ABERRANT HOMEOSTATIC T CELL
MIGRATION 54

NIK SIGNALLING IN DENDRITIC CELLS BUT NOT IN T CELLS IS REQUIRED FOR THE DEVELOPMENT OF EFFECTOR T CELLS AND CELL-MEDIATED RESPONSES 57

LOSS OF NIK FUNCTION RESULTS IN REDUCED T CELL PROLIFERATION, DIFFERENTIATION AND
PRODUCTION OF EFFECTOR CYTOKINES. 57
LOSS OF NIK RESULTS IN A PRIMARY APC DEFECT 60
DCS MATURATION AND CO-STIMULATION IS DEPENDENT ON NIK-SIGNALING 61
NIK$^{ALY/ALY}$ DCS ARE RESTRAINED IN T CELL PRIMING AND FAIL TO INDUCE EAE 64
RESTORED NIK-SIGNALING IN DCS BUT NOT IN T CELLS IS SUFFICIENT TO GENERATE PATHOGENIC T
CELLS 66
LOSS OF NIK-SIGNALING CRITICALLY IMPAIRS THYMIC DCS FUNCTION 68
RESTORATION OF NIK IN DCS RESCUES FOXP3, RORγT AND TBET EXPRESSION IN DEVELOPING
THYMOCYTES 70

DISCUSSION 73

THE ROLE OF SLTS FOR THE DEVELOPMENT OF CELL-MEDIATED IMMUNITY VERSUS HUMORAL
IMMUNE RESPONSES 75
CAN PRIMARY IMMUNE RESPONSES ONLY BE INITIATED IN SPECIALIZED LYMPHOID STRUCTURES? ... 75
HAS THE IMMUNODEFICIENCY IN NIK$^{ALY/ALY}$ MICE BEEN PROPERLY INTERPRETED? 76
IN CONTRAST TO T CELL ACTIVATION, B CELL ACTIVATION AND CLASS SWITCHING DEPENDS
STRICTLY ON INTACT LYMPHOID STRUCTURES 78
THE EVOLUTION OF LYMPHOID TISSUES AND THE IMPACT ON ADAPTIVE IMMUNITY 80
LYMPHOID STRUCTURES IN VERTEBRATE EVOLUTION 83
THE SPLEEN IN VERTEBRATE EVOLUTION 83
THE GUT-ASSOCIATED LYMPHOID TISSUES IN VERTEBRATE EVOLUTION 84
LYMPH NODES IN VERTEBRATE EVOLUTION 84
THE IMMUNE RESPONSES OF JAWED VERTEBRATES WITH A SPECIAL EMPHASIS ON COLD-BLOODED
VERTEBRATES 85
RESOURCEFUL IMMUNITY AND THE CONVERSION OF UNCONVENTIONAL LYMPHOID ORGANS 87
THE ROLE OF NIK-SIGNALING IN CELL-MEDIATED IMMUNITY 90

NIK-Signaling in T cell Function	90
NIK-Signaling in DC Function	91
NIK-Signaling in T cell Development	92
Does NIK-Signaling Equal Non-Canonical NFκB-Signaling?	93
NFκB-Signaling in Autoimmunity	94
Conclusion	**97**
Experimental Procedures	**99**
References	**105**
Acknowledgements	**119**

TABLE OF CONTENTS

ABBREVIATIONS

Ab	antibody
Ag	antigen
AID	activation-induced desaminase
Aly	alymphoplasia
APC	antigen presenting cell
BAFF	B cell-activating factor
BALT	bronchial-associated lymphoid tissue
BBB	blood brain barrier
BM	bone marrow
BMC	bone marrow chimera
BrdU	bromdesoxyuridin
CCL	chemokine ligand
CCR	chemokine receptor
CDR	complementary determining region
CFA	complete Freund adjuvant
CFSE	carboxyfluorescein succinimidyl ester
cIAP	cellular inhibitor of apoptosis
CMI	cell-mediated immunity
CNS	central nervous system
CTL	cytotoxic T lymphocyte
DC	dendritic cell
DTH	delayed type hypersensitivity
DTR	diphteria toxin receptor
DTx	diphteria toxin
EAE	experimental autoimmune encephalomyelitis
EBV	Eppstein-Barr virus
FCS	fetal calf serum
FDCs	follicular dendritic cell
GALT	gut-associated lymphoid tissue

GC	germinal center
GM-CSF	granulocyte/macrophage-colony stimulating factor
HEV	high endothelial venule
HLA	human leukocyte antigen
ICAM	inter-cellular adhesion molecule
Ig	immunoglobulin
IKK	inhibitor kappa B kinase
Il	interleukin
IFN	interferon
KLH	keyhole limpet hemocyanin
LN	lymph node
LPS	lipopolysaccharid
LT	lymphotoxin
MALT	mucosal-associated lymphoid tissue
MHC	major histocompatibility complex
MOG	myelin oligodendrocyte protein
MBP	myelin basic protein
MS	multiple sclerosis
mTEC	medullary thymic epithelial cells
NALT	nasal-associated lymphoid tissue
NFκB	nuclear factor kappa B
NIK	nuclear factor kappa B -inducing kinase
OVA	ovalbumin
PALS	periarteriolar lymphoid sheath
PAMPs	pathogen-associated molecular patterns
pDC	plasmacytoid dendritic cell
PELS	periellipsoid lymphoid sheath
PLT	primary lymphoid tissue
plt	paucity of LN T cells
PLP	proteolipid protein
PP	Peyers patch
PRR	pattern-recognition receptor
PTA	peripheral tissue antigens
PTx	pertussis toxin
Rag	recombination activating gene

RANK	receptor activator of NFκB
SEM	standard error of mean
SLT	secondary lymphoid tissue
TCR	T cell receptor
TGF-β	transforming growth factor beta
Th	helper T cell
TLR	toll-like receptor
TLT	tertiary lymphoid tissue
TNF	tumor necrosis factor
TNFR	tumor necrosis factor receptor
TRAF	TNF receptor-associated factor
Treg	regulatory T cell
TMEV	Theiler's murine encephalomyelitis virus
TWEAK	TNF-like weak inducer of apoptosis
VCAM	vascular cell adhesion protein
VLR	variable lymphocyte receptors
Wt	wildtype

ZUSAMMENFASSUNG

Die Vorstellung, dass sekundäre lymphatische Organe (SLOs) eine kritische Vorraussetzung für die Aktivierung von Lymphozyten sind, wird von der Beobachtung gestützt, dass alymphoplastische ($NIK^{aly/aly}$) Mäuse, die keine Lymphknoten oder Peyer Plaques besitzen, beeinträchtigte Immunität gegenüber Tumoren und Vakzinen aufweisen. Die strukturellen Fehlbildungen von $NIK^{aly/aly}$ Mäusen werden durch eine Mutation der NFkB-induzierenden Kinase (NIK) bedingt, welche den nicht-kanonischen NFkB-Signalweg aktiviert und Signale vom Lymphotoxin-β-Rezeptor transduziert. Letzterer ist an der Entwicklungsanlage von SLOs entscheidend beteiligt. Die Möglichkeit, dass NIK eine Rolle als Signalinstanz in der Immunaktivierung des ausgereiften Organismus spielt, wurde bisher vernachlässigt. Überraschenderweise haben wir entdeckt, dass die Immundefizienz von $NIK^{aly/aly}$ Mäusen teilweise in der Rolle von NIK in der T-Zellaktivierung begründet liegt und unabhängig von der Abwesenheit von SLOs ist. Während B-Zellaktivierung vom lymphoretikulären System und germinalen Zentren abhängt, um Antikörper-sekretierende Plasmazellen zu erzeugen, fungiert T-Zellimmunität komplett unabhängig von spezifischen SLO-Strukturen. Hiermit beschreiben wir ein alternatives Modell für die Induktion von zellvermittelter Immunität, bei der antigenpräsentierende Zellen Antigene aufnehmen und in die Leber migrieren, wo sie neolymphoide Aggregate bilden. Diese Strukturen reichen nicht aus um Affinitätsreifung und Klassenwechsel von B-Zellen zu ermöglichen, aber sie bieten eine Umgebung, die als Ersatz zur Erzeugung zellvermittelter Immunität dienen kann.

Auch wenn wir beschreiben, dass die Immundefizienz von $NIK^{aly/aly}$ Mäusen in unausreichender T-Zellaktivierung begründet liegt, zeigen wir nun, dass die NIK-Mutation in $NIK^{aly/aly}$ Mäusen nicht zu einem T-Zellintrinsischem Defekt führt, sondern zu einer Fehlfunktion in dendritischen Zellen, $CD4^+$ Effektorzelllinien zu kreieren. Die Expression von NIK spezifisch in dendritischen Zellen ist ausreichend, um zellvermittelte Immunität wiederherzustellen. Wir konnten jedoch zeigen, dass die pure Anwesenheit von NIK-suffizienten antigenpräsentierenden Zellen nicht genügt, um reife NIK-defiziente $CD4^+$ T-Zellen zu aktivieren. Vielmehr müssen NIK-defiziente $CD4^+$ T-Zellen eine T-Zellentwicklung im Thymus in Anwesenheit NIK-suffizienter dendritischer Zellen durchlaufen, um ein funktionales Effektorrepertoir zu bilden. Damit haben wir entdeckt, dass eine Population von thymischen dendritischen Zellen NIK benötigt, um T-Zelleffektorlinien während der thymischen Entwicklung zu bilden.

ZUSAMMENFASSUNG

SUMMARY

The notion that secondary lymphoid tissues (SLTs) are an absolute requirement for lymphocyte activation is supported by the finding that alymphoplastic (NIK$^{aly/aly}$) mice lacking lymph nodes (LNs) and peyer's patches display impaired immunity to tumors as well as vaccines. The developmental malformations of such NIK$^{aly/aly}$ mice are due to a mutation in the NFkB-inducing kinase (NIK), which activates the non-canonical NFkB pathway and is crucial for lymphotoxin-β receptor signaling. The latter is vitally involved in the formation of SLTs during development. Hence, the role of NIK as a signaling entity in immune activation in the mature organism has been somewhat overlooked. To our surprise, we found that the immunodeficiency of NIK$^{aly/aly}$ mice results from the impact of NIK on T cell priming but not from the lack of SLTs. We clearly demonstrate that while B cell activation requires a lymphoreticular system and germinal center (GC) formation for the generation of antibody-secreting plasma cells, against all conventional wisdom, T cell immunity appears to function completely independent of the structures provided by SLTs. We describe an alternative pathway for the induction of cell-mediated immunity (CMI), in which antigen-presenting cells (APCs) sample antigen and migrate into the liver where they induce neo-lymphoid aggregates. These structures are insufficient to support affinity maturation and class switching, but provide a novel surrogate environment for the initiation of CMI.

Although the immuno-deficiency in NIK$^{aly/aly}$ mice results from insufficient T cell priming, we further demonstrate that the NIK-lesion in NIK$^{aly/aly}$ mice does not cause an intrinsic CD4$^+$ T cell defect but leads to the failure of dendritic cells (DCs) to licence CD4$^+$ effector lineages. We discovered that DC-restricted expression of NIK is sufficient to restore effective CMI. However, adoptive transfer experiments have shown that the pure presence of NIK-sufficient APCs is not suitable to prime NIK-lesioned adult CD4$^+$ T cells to develop pathogenic effector properties. Yet, NIK-lesioned CD4$^+$ T cells need to undergo thymic development in the presence of NIK-sufficient DCs in order to shape the effector repertoire. By systematically studying the function of NIK in CMI, we discovered that a population of thymic DCs requires NIK and is capable to shape the formation of T effector lineages during development.

Summary

INTRODUCTION

THE IMMUNE SYSTEM

Pathogens such as viruses, bacteria, fungi and helminthes, constantly threaten the human body and ensure their survival by exploiting the host's resources. The immune system protects the body against diseases by attacking and eliminating invaders. Although the immune system is found in all creatures including plants, it has evolved to its highest complexity in mammals. It consists of different types of specific proteins, cells and organs that interact in a dynamic network.
In general terms, it can be divided into two categories: the innate and the adaptive immune system [1]. However, innate and adaptive immunity are bridged and depend on each other on several levels. The innate immune system is evolutionary ancient and recognizes components that are conserved among broad groups of pathogens. Due to its rapid activation it represents the first line of defense against invaders. Phagocytic cells such as macrophages, monocytes, dendritic cells and neutrophiles recognize patterns of the invader such as viral or bacterial DNA and cell wall components by specified receptors. Phagocytosis not only leads to the destruction of the pathogen, but also causes the release of inflammatory mediators that ultimately recruit other immune cells to the site of infection. However, the innate immune system is non-specific and does not lead to long-lasting immunity. In contrast, the adaptive immune system is pathogen-specific and has evolved only in vertebrates. It acts much slower than the innate immune system and is dependent on innate immune cells for activation. The cells of the adaptive immune system are known as thymus- and bone marrow-derived lymphocytes (T- and B cells, respectively). They carry receptors, which are generated by somatic recombination and recognize components that are specific to the pathogen, so-called antigens. B cells that recognize the antigen via a membrane-bound immunoglobulin (B cell receptor) will produce and secrete antigen-specific antibodies. Antibodies can either directly neutralize the antigen, such as bacterial toxins, or induce secondary immune functions such as the activation of the complement system, an association of small proteins that lead to the lysis of infected cells via a membrane attack complex. T cells can be divided into different subsets that are phenotypically distinct in the expression of clusters of differentiation (CD), but more importantly play functionally discrete roles. $CD8^+$ and $CD4^+$ T cells are characterized by the expression of a T cell receptor (TCR) that consists of one α and one β chain. While $CD8^+$ T cells have cytotoxic activity and can destroy infected cells and tissues, $CD4^+$ T cells provide help to B cells and secrete cytokines that attract and activate other immune cells. Another minor T cell subset carries a TCR

consisting of one γ and one δ chain (γδ T cells). These T cells are often classified as part of the innate immune system due to that the specificity of their TCR often is restricted to common molecules produced by microbes. γδ T cells are predominantly found in the gut but also occur in other organs such as in the skin.

The evolutionary advantage of the adaptive immune system compared to the innate immune system is that it has the capacity to "remember". The immunological memory is based on the preexistence of a clonally expanded population of antigen-specific lymphocytes that allows a more rapid and efficient action when the organism repetitively encounters the same pathogen. In the course of evolution, the appearance of adaptive immunity coincided with the development and complexification of lymphoid organs. These organs provide a microenvironment in which cells of the innate and adaptive immune system can meet and interact efficiently. Lymphoid structures are believed to be crucial for the induction of both B cell-driven (humoral) and T cell-driven (cell-mediated) immunity. This thesis will mainly focus on the structural and molecular requirements of cell-mediated immunity (CMI) using a model of T cell-driven autoimmunity. The following chapters will introduce the structure and function of lymphoid organs in more detail. Subsequently, different aspects (functional and molecular) of T cell development, activation as well as tolerance and autoimmunity will be discussed. Introducing important signaling pathways as well as the mouse disease model used in this work will follow in order to provide the reader with fundamental knowledge to understand the work presented in this thesis.

LYMPHOID TISSUES: WHERE T CELLS DEVELOP AND POLARIZE

Three types of lymphoid tissues can be distinguished. While primary lymphoid tissues are a place of hematopoiesis and development of lymphocytes, secondary and tertiary lymphoid tissues are places where immune cells become activated to convert into effector cells. The following chapter will explain both structure and their function, specifically in cell-mediated immunity, of these lymphoid tissues.

PRIMARY LYMPHOID TISSUES

Bone marrow and thymus are the classical primary lymphoid tissues (PLTs). The bone marrow is the site where hematopoietic multipotent stem cells give rise to all blood cell types including myeloid and lymphoid lineages. Myeloid precursors develop into macrophages, neutrophiles, basophiles, eosinophiles, erythrocytes, megakaryocytes and dendritic cells (DCs). Lymphoid precursors give rise to T cells and B cells as well as natural killer (NK) cells. While B cell development occurs within the bone marrow, T cell development occurs within the thymus. Therefore, T cell progenitors migrate from the bone marrow to the thymus where they undergo clonal expansion and several selection processes, until they can leave the thymus as mature T cells.

THYMIC T CELL DEVELOPMENT

The thymus is a tissue without any self-renewing potential and is therefore depending on the immigration of new progenitors from blood and bone marrow [2]. It is a two-lobed organ situated just below the upper end of the sternum and can be morphologically divided into the outer cortex and the inner medulla, through which within approximately 20 days thymocytes migrate in a strictly organized fashion during their development (Figure 1). The earliest precursors that enter the thymus derive from hematopoietic stem cells in the bone marrow and are known as early T cell progenitors. These cells home to the thymus through the cortico-medullary junction and are not yet T lineage committed. They do not yet express CD4 or CD8 and are termed double negative cells (DNs). DN thymocytes can be divided into four sequential phenotypic stages (DN1, DN2, DN3 and DN4), which are defined by different expression levels of CD44 and CD25 [3]. During all DN stages, thymocytes proliferate, expand and migrate through the outer cortex, where they interact with cortical thymic epithelial cells. T lineage commitment occurs during the DN2 stage and is largely dependent on Notch1 signaling as well as other transcription factors. The rearrangement of the variable V, D and J gene segments of the TCR β-chain locus occurs within the DN3 stage. Because recombination activating gene (Rag)-mediated recombination often leads to rearrangements that are out of frame, cells have to pass a checkpoint called β-selection. In order to survive, thymocytes need to express a functional pre-TCR consisting of a properly rearranged TCRβ chain, CD3 and the invariant pre-Tα-chain. Cell death will be induced in cells that fail to successfully rearrange their TCRβ locus. Thymocytes that pass β-selection will enter another proliferative burst before they initiate the expression of both CD4 and CD8, thus entering the double positive (DP) stage. DPs rearrange the V and J segments of the TCRα-chain and consequently undergo the next checkpoint: positive selection. During positive selection, the fully rearranged α/β-TCR is examined for its capability to interact with major histocompatibility complexes (MHCs). Only thymocytes that bind MHC/antigen complexes with adequate affinity will receive further survival signals. Thymocytes that do not recognize MHC/antigen complexes will die through apoptosis (death by neglect). Approximately 5-10% of DPs will survive positive selection [4, 5]. Furthermore, the process of positive selection will determine the lineage of the T cell. DPs that are positively selected on

MHCII complexes will become CD4$^+$ T cells, while DPs recognizing antigens on MHCI complexes will mature into CD8$^+$ T cells. Single positive (SP) thymocytes will now migrate to the medulla, where they have to pass another checkpoint before they can be released into the periphery. Because positive selection is based on the random recombination of the TCR and the recognition of self-antigens presented on MHC complexes, this process bears the potential for autoimmune reactions. Therefore, a second process has evolved to prevent the maturation of T cells with overt reactivity towards self-antigens, namely negative selection. Here, TCR-bearing thymocytes with a high affinity for MHC-self antigen complexes will undergo programmed cell death. Medullary thymic epithelial cells (mTECs) play an important role in negative selection. They express and present multiple peripheral antigens to the developing thymocytes, a process that depends on the transcription factor autoimmune regulator (AIRE). Negative selection is an important component of immunological tolerance and aims to prevent the formation of auto-reactive T cells that could potentially induce autoimmunity. However, this selection process is not 100% complete and some auto-reactive thymocytes will exit the thymus as mature T cells. Consequently other mechanisms of tolerance induction have been established and will be discussed later on.

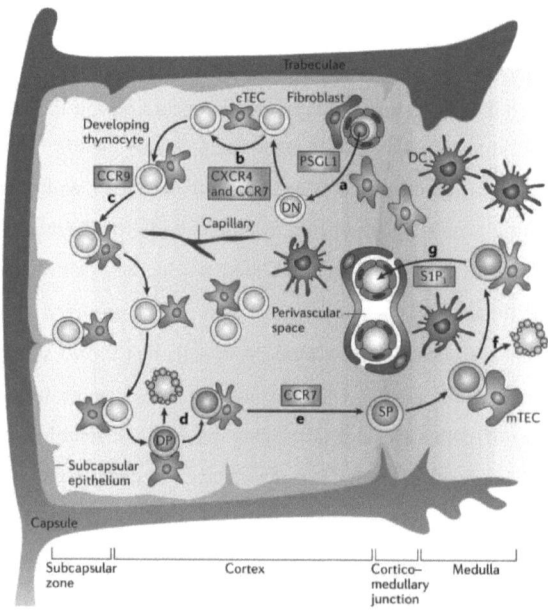

Figure 1: T cell development and selection (adapted from [6]). T cell progenitors enter the thymus at the cortico-medullary junction **(a)**. DN thymocytes migrate through the outer cortex. During that process they interact with cortical epithelial cells, undergo β-selection and mature to DP thymocytes **(b and c)**. DPs undergo positive selection and T cell lineage commitment **(d)**, before they migrate as SPs into the thymic medulla **(e)**. There, SPs interact with medullary epithelial cells and undergo negative selection **(f)**, before they leave the thymus as mature T cells **(g)**.

Secondary Lymphoid Tissues

A human body has one spleen, approximated 500-700 lymph nodes (LNs), many of them hardly visible, and around 30 Peyer's patches (PP). All these organs as well as tonsils, adenoids and nasal- or bronchus-associated lymphoid tissues (NALT or BALT) belong to the category of secondary lymphoid tissues (SLTs) [7]. SLTs enhance the efficiency of immune responses by arranging B and T cells in anatomical locations that favor their interaction with antigen-presenting cells (APCs). Furthermore, they provide a framework, which allows rapid circulation of naïve cells through a location where antigens are concentrated [8].

The spleen consists of a white pulp and a red pulp, both executing very distinct functions. The white pulp is composed of lymphoid follicles that are rich in B cells and periarteriolar lymphoid sheaths, in which mainly T cells are located. The red pulp is responsible for clearing the body from senescent erythrocytes, particles, certain encapsulated bacteria and protozoa. The marginal zone separates the red and the white pulp. The spleen drains predominantly blood-borne antigens, which are picked up by professional APCs that in turn activate humoral and cell-mediated immune responses.

LNs are located at vascular junctions and are connected to lymphatic vessels that deliver antigens and APCs. They consist of a cortex and a medulla (Figure 2 A). B cells are mainly arranged in follicles within the outer cortex and have the potential, as in the spleen, to form germinal centers (GC). GCs are structures in which B cell affinity maturation and antibody class switch takes place. T cells are predominantly located within the inner cortex. Antibody-secreting B cells (plasma cells) can be found within the medulla. LNs are connected to the lymphatic system and filter lymph rather than blood. The lymph delivers antigens and APCs that sample antigens in peripheral tissues to the node.

PPs are aggregations of follicles populated by B and T cells that bulge into the gut lumen. PPs are not connected to afferent lymphatics and obtain their antigens from the lumen of the mucosa via so-called microfold (M) cells [9]. The function of PPs is the immuno-surveillance of the intestinal lumen.

INTRODUCTION

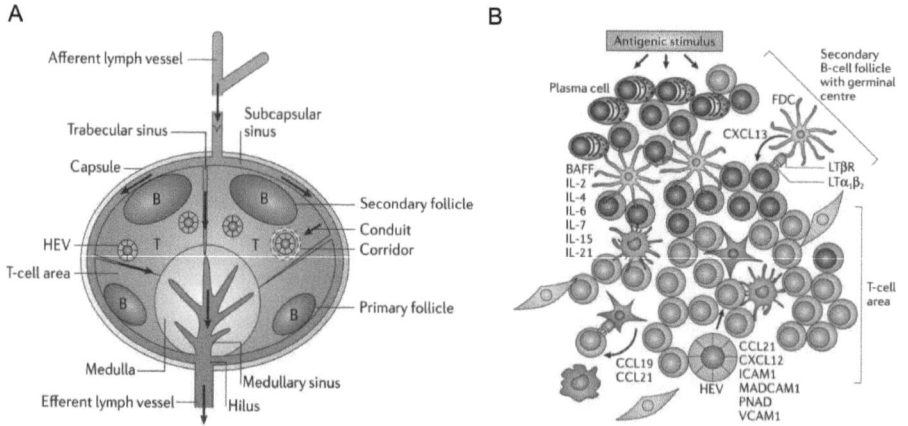

Figure 2: Basic structure of secondary and tertiary lymphoid tissue (adapted from [10]). (A) The main structural components of lymph nodes are shown. (B) Lymphoid neogenesis upon chronic inflammation leads to tertiary lymphoid tissues.

T CELL ACTIVATION AND POLARIZATION

Naïve T cells circulate through SLTs to scan for MHC complexes that present the antigen specific for their TCR. Professional APCs, such as macrophages and DCs, are innate immune cells that are activated by microbial components via pattern-recognition receptors (PRRs) e.g. Toll-like receptors (TLRs). The T cell must acquire three signals provided by the APC in order to become sufficiently activated (Figure 3). The first signal is the recognition of the MHC/antigen complex by the antigen-specific TCR. The second signal stabilizes the immunologic synapse via recruitment of co-stimulatory adhesion molecules present on both APC and T cell, e.g. CD80/CD86 on APCs and CD28 on T cells. The third signal is comprised via the secretion of cytokines by APCs, which signal via cytokine receptors on T cells. These cytokines polarize T cells toward a certain effector phenotype [11]. The activation of an antigen-specific T cell via these three signals leads to its massive clonal expansion and a differential gene expression that will result in various T cell phenotypes.

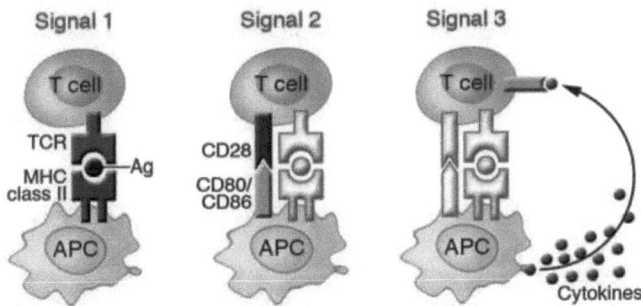

Figure 3: The 3 signals required for T cell activation (adapted from [11]). Signal 1 consists of the binding of the TCR to the MHC/antigen complex. **Signal 2** stabilizes the immunological synapse by binding of the co-stimulatory molecules CD80/CD86 to CD28. **Signal 3** polarizes the T cell into a specific T cell phenotype via the secretion of specific cytokines by the APC.

In 1986 CD4$^+$ helper T cells were divided into two independent subtypes, namely Th1 and Th2 cells [12]. While the pro-inflammatory cytokines interleukin-12 (Il-12) and interferon gamma (IFNγ) lead to the differentiation of Th1 cells that express the transcription factor Tbet, Il-4 induces the polarization of GATA3-expressing Th2 cells. Th1 cells secrete Il-2, TNFα and large amounts of IFNγ. They are responsible for CMI, defense against intracellular pathogens and promote the differentiation of cytotoxic CD8$^+$ T cells. IFNγ enhances the phagocytic activity of macrophages and further up-regulates MHCII molecules on APCs. On the other hand, Th2 cells produce mainly Il-4, Il-5, Il-10 and Il-13. They stimulate humoral immunity by aiding B cell activation and antibody class switch. Th2 cells mediate allergic reactions and anti-helminth responses.

Within the last decade many additional CD4$^+$ T cell subtypes have been described (Figure 4) [13, 14]. Thymus-derived, natural regulatory T cells (nT$_{regs}$) express the transcription factor forkhead box P3 (FoxP3) and secrete transforming growth factor beta (TGF-β), Il-10 and Il-35. In the periphery, regulatory T cells that also express FoxP3 can be induced (iT$_{regs}$) in the presence of TGF-β or retinoic acid (RA). A third population of regulatory T cells (Tr1), though they lack the expression of FoxP3 but produce high quantities of Il-10, has also been described [15]. Follicular helper T cells (T$_{FH}$) can develop in the presence of Il-6 and Il-21 [16]. Another T helper subtype that was recently described and strongly associated with different autoimmune diseases is the Th17 lineage. Th17 cells are induced by TGF-β, Il-6 and Il-23 and secrete Il-17A, Il-17F and Il-22 and express the transcription factor retinoic acid-related orphan receptor gammat (RORγt) [17, 18]. A combination of Il-4 and TGF-β has been shown to induce CD4$^+$ T cells that secrete Il-9, and these cells have recently been named Th9 cells [19, 20].

INTRODUCTION

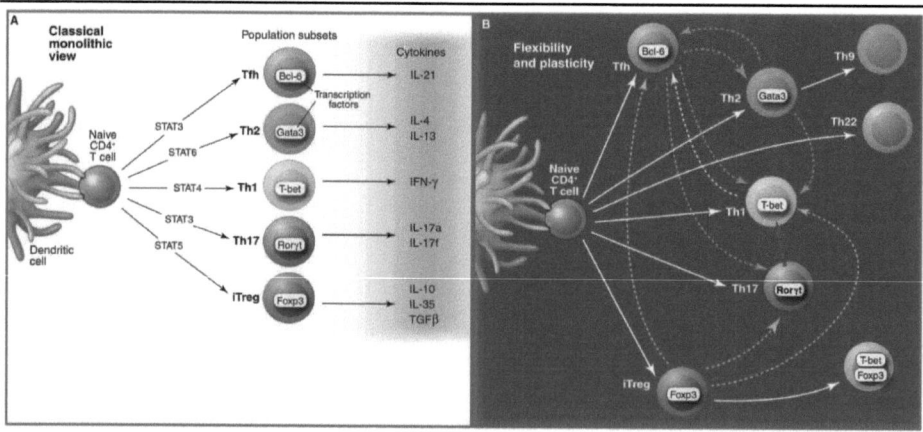

Figure 4: T helper cell differentiation and plasticity, adapted from [14]. (A) The classical monolithic view is based on the original idea that T cell subsets behave like inflexible lineages with the expression of lineage-defining transcription factors and a certain cytokine profile. **(B)** Recent studies revealed that T cell lineages are flexible and can be converted into other T cell subsets dependent on the present cytokine microenvironment. In the process of plasticity, T cell subsets can change their cytokine profile and the expression of master regulators and transcription factors.

Classically, these $CD4^+$ T cell subsets have been thought to be stable T cell lineages characterized by the expression of a certain cytokine profile as well as their master transcription factors (Figure 4A). Recent studies suggest, that T cell phenotypes are more plastic than once thought (Figure 4B). For example TGF-β has been shown to reprogram Th2 cells by repressing GATA3 and Il-4, but inducing the production of Il-9 [19]. T_{regs} can be converted to Th17 cells and into IFNγ-producing cells. Th17 cells in the absence of TGF-β and in the presence of Il-12 can transit into Th1 cells [21]. One recent report has demonstrated that Th17 cells, that develop in the absence of TGF-β but in the presence of Il-23 mature into autoimmune pathogenic T cells that express both transcription factors Tbet and RORγt [22]. The accumulation of data describing evidence for T cell plasticity revolutionizes our understanding of the different T cell subsets.

TERTIARY LYMPHOID TISSUES

As shown above, PLTs and SLTs are clearly defined organs. Their location within the body is determined during ontogenesis. In contrast, tertiary lymphoid tissues (TLTs) are not genetically fixed and lack a capsule. They often show lymphoid tissue-specific organization with T cell zones and B cell follicles that contain follicular dendritic cells (FDCs) (Figure 2B). TLTs develop ectopically and are induced by chronic inflammation, autoimmunity or chronic graft rejection [7].

The neogenesis of TLTs has been observed in several organs: lungs, joints, thyroid, liver and CNS to only name a few [23].

SELF TOLERANCE AND AUTOIMMUNITY

The activation of the antigen-specific (adaptive) immune system is initiated within SLTs or TLTs by the recognition of antigens in association with so-called "danger signals". These danger signals are usually components of invaders such as bacteria, parasites, worms or viruses that signal our body an infection or the like. Followed by that, antigen-specific activation of either T or B cells will lead to clearance of the pathogen. If immune cells within a healthy body recognize their antigen in the absence of danger signals, tolerance is induced. Therewith, one of the main principles of the adaptive immune system is to tolerate *self*-antigens (components of our body) and only attack when the body is challenged by pathogens. Immunological tolerance comprises two different mechanisms, namely central tolerance and peripheral tolerance. In central tolerance, developing lymphocytes (T cells in the thymus and B cells in the bone marrow) are scanned for their auto-reactivity early during development and consequently eliminated via apoptosis. Peripheral tolerance can be induced on different levels: via anergy, deletion or suppression by regulatory T cells [24].

Autoimmunity is defined as a failure of the immune system to tolerate *self*-antigens of the body. The immune system recognizes *self*-antigens of the body as *foreign*, which leads to an attack of cells and whole tissues. How this "confusion" of the immune system to discriminate *self* from *foreign* is initiated, is up to today poorly understood. Possible triggers are molecular mimicry by viral components or environmental factors. Autoimmune diseases affect approximately 5% of the population in industrial countries. Prominent examples are systemic lupus erythematosus, rheumatoid arthritis, Crohn`s disease, diabetes mellitus type 1, psoriasis, ulcerative colitis and multiple sclerosis [25].

In this thesis, a model of T cell-driven encephalomyelitis that resembles clinical aspects of multiple sclerosis is used in several experimental setups. The following chapter will introduce this autoimmune disease and its animal model in more detail.

Multiple Sclerosis

Pathology and Symptoms

Multiple sclerosis (MS) is a demyelinating disease of the central nervous system (CNS), which is often progressive and results in neurological damage and disability [26]. MS is considered a classical T cell-mediated autoimmune disease, that leads to degenerative processes including inflammatory injury of axons and glia, diffuse damage to white matter and involvement of both deep and cortical gray matter [27]. A hallmark of MS is the formation of lesions and plaques within the white matter. These are generally areas of inflammation, demyelination and gliosis. Myelin and oligodendrocyte auto-reactive T cells release toxic inflammatory mediators within the brain that result in axonal and glial injury. This inflammation in turn leads to the activation of astrocytes and the release of growth factors that encourage remyelination and partly functional recovery. The neurological symptoms lead to both physical and cognitive disabilities [26]. Physical disabilities include general motor impairments, tremor, poor balance, stiffness, painful spasm, and impaired speech and swallowing. Cognitive impairments appear in deficits in attention, reasoning and fatigue. The most common form of the disease is the relapsing-remitting MS, which is characterized by phases of immune attack and destruction that are followed by phases of remyelination and recovery. Relapsing-remitting MS can develop into a secondary progressive MS. A small percentage of patients experience a form of the disease known as primary progressive MS, where no phases of recovery occur.

Epidemiology

MS affects around 2 million people worldwide and is the most common non-traumatic, disabling neurological disease of young adults [28]. Areas displaying the highest levels of prevalence are northern and central Europe, northern USA, Canada, southeastern Australia, the former Soviet Union and New Zealand, suggesting that environmental factors found pre-dominantly within industrial countries might partly trigger the disease [29, 30].

Risk Factors

MS results from the interplay between yet unidentified environmental factors and susceptibility genes. The most prominent association of genetic risk and MS has been described for human leukocyte antigen (HLA) class II molecules [31]. HLA types positively or negatively associated with the disease are region-specific. The two HLA-DRB genes of the HLA-DR15 haplotype confer the largest part of genetic risk in Caucasians of Western Europeans [32]. Also other genes, though with more modest effects, have been associated with susceptibility for the disease, such as interleukin-7 receptor α (Il-7RA) and Il-2RA, C-type lectin-domain family 16 member A (CLEC16A), CD58, tumour-necrosis-factor receptor superfamily member 1A (TNFRSF1A), interferon regulatory factor 8 (IRF8) and CD6 [33]. Furthermore, ethnic origin, most likely based on genetic determinants, has been associated with disease susceptibility. Curiously, the disease is virtually absent in Chinese and Filipinos [34]. Although MS is more common in women than in men (2:1) [35], no disease-related genes have been identified on the X chromosome, which suggests that the sex-related preference of MS might be associated with female-specific physiology (e.g. hormones) [36].

Environmental risk factors associated with the disease include Epstein-Barr virus (EBV), smoking and vitamin D [31]. Although EBV is a ubiquitously latent infection of our population with 94% incidence, 99% of MS patients are EBV positive [37]. People that have experienced infectious mononucleosis have an increased risk to develop MS as well as people with high titers of anti-EBV antibodies [38].

Therapies

Most therapeutical approaches aim at disabling the immune system, which can be achieved in several ways: (i) Whole immune cell populations, that are thought to be involved in the pathogenicity, can be targeted. (ii) The migration of lymphocytes from the periphery to the CNS can be blocked. Alternatively, (iii) the most direct approach lies in "resetting" the immune system by wiping out the existing immune repertoire (including auto-reactive cells) and allowing a healthy immune system to regenerate [28]. The drawback of all these strategies is that they also deplete "normally" functional immune cells and therewith compromise general immune protection.

The following paragraph will briefly summarize current drugs that are used in MS treatment:

Glatiramer acetate (GA) is a heterogenous mixture of proteins and polypeptides that acts as an altered peptide ligand for myelin basic protein [39]. The induced GA-specific T cell clones tend to express predominantly a Th2 phenotype, which produces anti-inflammatory cytokines such as Il-4, Il-10 and TGF-β. These GA-specific Th2 cells can then be found within the CNS of MS patients, presumably causing a phenomenon known as "bystander suppression" of inflammation. GA furthermore also exhibits neuroprotective activity by inducing the secretion of neurotrophic factors such as brain-derived neurotrophic factor (BDNF) and neurotrophin 3 and 4 in the brain.

Interferon-beta is one of the oldest and most classical treatments in MS [40]. Like GA, it belongs to a drug category that exerts anti-inflammatory and immuno-modulatory effects. Interferon-beta inhibits MS disease activity by on one hand inhibiting pro-inflammatory cytokines, nitric oxide synthase, T cell activation and migration and on the other hand stimulating the production of anti-inflammatory cytokines such as Il-10 and Il-4.

Natalizumab is a humanized IgG4 monoclonal antibody that binds very late antigen-4 (VLA-4). VLA-4 is an adhesion molecule expressed on lymphocytes that binds to vascular cell adhesion molecule-1 (VCAM-1), which is expressed by cerebrovascular endothelial cells. It mediates the migration of lymphocytes through the blood-brain barrier (BBB) into the CNS. Blocking this process prevents the infiltration of the CNS with auto-reactive immune cells.

Fingolimod (FTY720) acts as a super-agonist to the sphingosine-1-phosphate receptor (S1PR), which is expressed by lymphocytes and mediates their migration into and out of SLTs. Blocking this interaction leads to the sequestering of lymphocytes within SLTs and prevents the migration to the CNS.

Daclizumab is a humanized IgG1 antibody that binds CD25. The intention of this drug was to specifically target activated T cells. However, Daclizumab was shown to expand a subset of NK cells that express high levels of CD56 and state immuno-regulatory properties by inhibiting the survival of activated T cells [41].

Rituximab is a chimeric murine/human IgG1k CD20-specific antibody, which triggers cytolysis in pre-B cells. The success of a phase II trial using rituximab supports the idea that B cells play a role in the pathology of the disease. However, because rituximab does not target plasma cells, it is unlikely that its effect is conducted via the reduction of pathogenic autoantibodies but rather by modulating antigen-presentation and cytokine production that regulate T cell responses [42].

Alemtuzumab (Campath-1H) is a humanized anti-CD52 antibody, which leads to massive lymphopenia, and disables the immune response by simply removing lymphocytes. CD52 is expressed on the surface of lymphocytes and monocytes, but its function is unknown. After treatment, $CD4^+$ T cells need up to 5 years to recover to pre-treatment levels. Alemtuzumab is also used as treatment for other disease, such as chronic lymphocytic leukemia, renal transplantation and a variety of other autoimmune diseases [43-45].

Cladribine (2-chlorodeoxyadenosine) is an adenosine deaminase-resistant nucleoside analogue with selective lymphotoxicity. It incorporates into the DNA, causing DNA damage and cell death. It thereby follows the principle of *alemtuzumab* treatment by inducing lymphopenia.

EXPERIMENTAL AUTOIMMUNE ENCEPHALOMYELITIS (EAE)

EAE was established over 50 years ago [46]. It is used as an animal model for autoimmune inflammation of the CNS, which resembles many characteristics also observed in MS. EAE is a T cell-driven autoimmune disease and can be induced by subcutaneous (s.c.) immunization of animals with myelin proteins or antigens that are emulsified in complete Freund`s adjuvant (CFA). Myelin proteins/antigens used for immunization are e.g. myelin basic protein (MBP), proteolipid protein (PLP) or myelin oligodendrocyte glycoprotein (MOG). Furthermore, the animals are treated with pertussis toxin (PTx), which has been shown to facilitate the migration of immune cells into the brain. PTx also contributes to breaking T cell tolerance and promoting clonal expansion and cytokine production by T cells [47, 48]. In murine EAE, the disease manifests by the paralysis of tail and hind limbs, which then progresses to the forelimbs. The area of inflammation is mostly targeted to the spinal cord and the cerebellum [49].

Besides active immunization with myelin proteins, EAE can also be induced via different strategies. Classical passive immunization is based on the adoptive transfer of previously activated myelin-specific $CD4^+$ T cells into naïve mice. Alternatively, also the adoptive transfer of MOG- or MBP-specific $CD8^+$ T cells can mediate severe CNS autoimmunity [50, 51]. Certain virus models are capable of inducing encephalomyelitis and are therefore useful tools in understanding the potential viral etiology of MS. The exposure to Theiler`s murine encephalomyelitis virus (TMEV) can for example lead to fatal encephalomyelitis and immune-mediated demyelination [52, 53]. Transgenic mouse models that lead to the ablation of brain-specific cell types allow the investigation of demyelinating processes that are not initiated in the periphery of the host but rather within the CNS. The depletion of oligodendrocytes via a diptheria toxin-mediated cre-sprecific mouse model results in severe myelin loss that leads to tremor, hind limb paralysis and weight loss [54]. Furthermore, toxin-based models like cuprizone, lysolecithin, or lipoploysaccharide (LPS)-injections into the brain are used to study the general mechanisms of de- and remyelination [55-57].

Altogether, these different models serve to understand general aspects of MS. However, there are limitations and cautiousness that have to be kept when interpreting data coming from animal models. The initial trigger of MS remains to be identified. For that reason, drug testing and clinical trials cannot yet be replaced by the study of the above-described models.

We have learned above that the thymus will eliminate T cells that carry TCRs with a high affinity to self-antigens. However, despite this selection process, self-reactive T cells can be found in our body. These ones need to be tightly controlled within the periphery. For instance, an auto-reactive T cell will be rendered tolerant, if one of the three signals provided by APCs is missing. In the model of EAE, T cells are not only presented with an auto-antigen, but an adjuvant consisting of microbiological components will deliver the additional danger signal, that activates APCs to provide all three signals and therewith breaks tolerance in auto-reactive T cells.

NFκB-SIGNALING

Besides the structure of lymphoid tissues and the three activating signals provided by APCs, all immune cells depend on molecular signaling pathways that allows them to translate incoming receptor signals into a differential gene expression. One of the ubiquitous signaling cascades frequently used by all immune cells is the nuclear factor kappa B (NFκB) pathway. For the work presented in this thesis, not only the requirements of SLTs in CMI are of interest, but also molecular requirements, in particular these of a specific kind of NFκB-signaling. The following chapter will therefore introduce this signaling cascade in more detail.

NFκB regulates genes that are involved in various stages of the immune response such as activation of innate cells and lymphocytes, DC maturation, inflammation and survival. NFκB represents a family of transcription factors, which in mammals comprises of RelA (p65), RelB, c-Rel, NFκB1 (p50) and NFκB2 (p52) [58]. These are structurally homologous proteins that form homo- or heterodimers via their terminal Rel homology domains (RHDs). In unstimulated cells the RHDs of NFκBs associate with inhibitory proteins of κB (IκBs) and are sequestered in the cytoplasm. Both NFκB1 and NFκB2 comprise of large precursor proteins (p105 and p100, respectively) that need to be post-translationally processed (to p50 and p52, respectively) for their activation. While p105 processing is constitutive, the processing of p100 is regulated [59]. Most NFκB dimers are activators, although p50/p50 and p52/p52 homodimers are repressive in gene activation.

Generally, NFκB can be divided into two pathways, the canonical and the non-canonical (also classified as classical and alternative, respectively), which will be discussed in more detail below (Figure 5).

INTRODUCTION

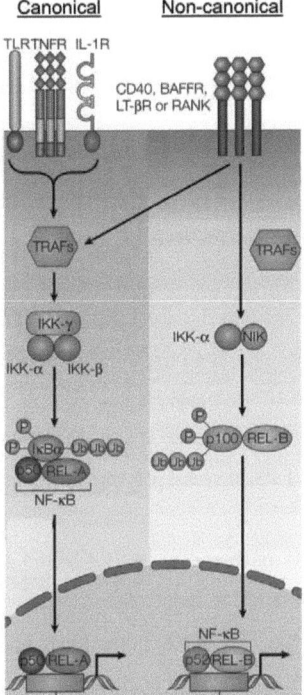

Figure 5: Two NFκB-pathways (adapted by [60]). TNFR family members, Il-1R and TLRs are inducing canonical NFκB-signaling, whereas CD40, BAFFR, LTβR and RANK induce non-canonical NFκB-signaling.

CANONICAL NFκB-SIGNALING

Pro-inflammatory cytokines and pathogen-associated molecular patterns (PAMPs) cause, via different tumor necrosis factor receptors (TNFRs) or TLRs, activation of the IκB kinase complex (IKK). The most prominent IKK complex is the so-called NEMO complex, consisting of two catalytic (IKKα and IKKβ) and one regulatory subunit (IKKγ). The IKK complex catalyzes the phosphorylation of IκB, its polyubiquitination and subsequent degradation by the 26S proteasome. The released NFκB dimers translocate to the nucleus and activate gene transcription [61].

The knowledge obtained of the physiological function of canonical NFκB-signaling stems largely from the analysis of NFκB1$^{-/-}$ mice. These mice show multifocal defects in immune system function [62]. Data from NFκB1$^{-/-}$ mice have revealed that the maturation and survival of B cells as well as the humoral response to T cell-dependent antigens depends on NFκB1. CD4$^+$ T cells depend on canonical NFκB-signaling for their proliferation and up-regulation of CD25, as well as their functional maturation into different T helper phenotypes. Macrophages depend on NFκB1 for the production of Il-6 and Il-12 upon encountering LPS stimulation. The response and control to certain infectious organisms relies on canonical NFκB-signaling, including *Leishmania major* and *Streptococcus pneumonia* infections [62, 63].

Non-Canonical NFκB-Signaling and NIK

For many years it has been believed that the non-canonical NFκB-pathway is preferably activated by ligands important for either the lymphoid organogenesis (through LTβ) or in B cell responses (through CD40L and BAFFL) [64]. However, it has become increasingly evident that the non-canonical NFκB-pathway can be triggered by many different ligands such as RANK, LIGHT, TWEAK, CD70, CD28, Il-1 and LPS, thereby playing an important role in various responses [65-72].

Non-canonical NFκB-signaling is independent of IKKβ and IKKγ [73]. Instead it requires the NFκB-inducing kinase (NIK) and IKKα to induce the slow processing of p100 to p52. NIK directly phosphorylates and activates IKKα. An IKKα homodimer phosphorylates the precursor protein p100 that leads to its polyubiquitination and degradation to the active form p52. Active p52 most commonly dimerizes with RelB [74].

The activity of NIK is tightly regulated on several levels [75]. In resting cells, NIK is constitutively degraded through ubiquitination by a complex consisting of TNF receptor-associated factors (TRAF) 2 and TRAF3 and cytosolic inhibitor of apoptosis (cIAP) 1 and cIAP2 [76, 77] thereby preventing basal activation of this pathway (Figure 6A). The signal-induced activation of the non-canonical pathway results in the degradation of TRAF2 and TRAF3, leading to the stabilization of NIK protein [78]. A second control level has recently been described for the regulation of NIK activity in stimulated cells where NIK and IKKα form a specific kinase complex, in which NIK phosphorylates IKKα and recruits it to p100 [79]. In that process IKKα also phosphorylates NIK and targets it for degradation [80], thereby controlling the magnitude of NIK accumulation via a negative-feedback mechanism (Figure 6B).

INTRODUCTION

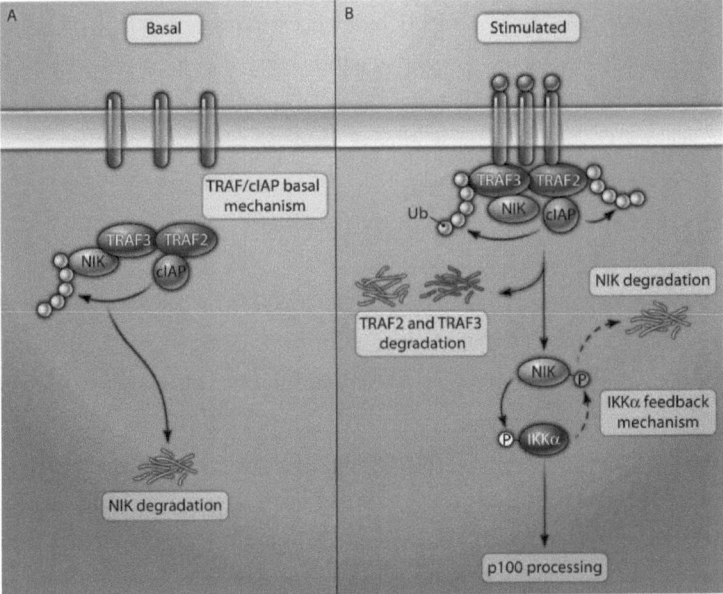

Figure 6: Regulation of NIK activity (adapted by [75]). (A) In unstimulated cells, cIAP1 and cIAP 2 constantly target NIK for ubiquitination and degradation via interaction with TRAF2 and TRAF3. **(B)** Upon activation, TRAF 2 and TRAF 3 are targeted for cIAP-mediated ubiquitination and degradation by the stimulated receptor complex. NIK is released and stabilized and can activate IKKα, which triggers the processing of p100. In turn, activated IKKα phosphorylates and induces the degradation of NIK to prevent an overaccumulation of NIK in activated cells.

There is evidence that NIK also signals into the classical NFκB-pathway. For example, CD40, BAFF and CD27-signaling depend on NIK for the activation of both pathways [67]. Also the inhibition of TRAF3, the essential negative regulator of the non-canonical pathway, results in activation of both canonical and non-canonical signaling cascades [81]. A recent report describes that LTβR-induced signaling via NIK can result in the phosphorylation of p65, which together with RelB synergistically enhances the gene expression of granulocyte-macrophage stimulating factor (GM-CSF), a cytokine that is produced by T cells and plays a role in DC development [82]. These reports challenge the prevailing wisdom of two distinct NFκB-pathways and it is important to further investigate the circumstances under which this cross-talk can occur.

NFκB AND NIK PLAY A ROLE IN VARIOUS DISEASES

NFκB plays an important role in the initiation and promotion of cancer by fostering an inflammatory milieu in which various cytokines aid and abet malignant transformation [83]. Deregulated activation of NFκB contributes to diseases that range from chronic inflammation and autoimmunity to cancer [84-86]. In a variety of lymphoid cancers, NFκB is constitutively active due to diverse somatic mutations, genomic amplifications or chromosomal translocations. For example, mucosa-associated lymphoid tissue (MALT) lymphoma can be induced by a genetic translocation that leads to a fusion protein of cIAP2 and MALT, which stabilizes the activation of the canonical NFκB-pathway. Mutations in NIK that interrupt the TRAF3-binding domain can cause highly stable NIK proteins that have been associated with the development of multiple myeloma [87]. Also high-level amplifications of the NIK genomic locus and NIK gene translocations that lead to overexpression, as well as TRAF3 gene deletions or deletions affecting the closely linked cIAP1 and cIAP2 have been identified in multiple myeloma [83, 85].

Chronic activation of NFκB due to persistent viral infection can also promote the progression to malignancies [88]. The latent membrane protein 1 (LMP1) of EBV can induce NIK-dependent p100 processing that has been associated with EBV-positive Hodgekin's lymphoma [89-91]. The Kaposi Sarcoma-associated herpesvirus (KSHV) can induce primary effusion lymphoma through the viral protein vFLIP that can lead to IKK activation. The human T cell lymphoma virus (HTLV) expresses the tax oncoprotein that binds to NEMO and thereby induces IKK activation or alternatively activates the IKKα-dependent non-canonical NFκB-pathway [85].

Although the contribution of NFκB-signaling seems to mainly add to lymphoid malignancies, there is evidence that the transformation of solid cancers, such as breast cancer, is effected by dysregulated NFκB [92].

Taken together, NIK poses an attractive pharmacological target for the treatment of a variety of diseases [83] and it is thus important that its role and function within the immune system is resolved.

ALYMPHOPLASIA MICE

As mentioned above, non-canonical NFκB-signaling is a prerequisite for the *anlage* of SLTs. Mice carrying lesions in elements of this pathway are often alymphoplastic (absence of LNs) and lack the specific lymphoid organization in spleen and thymus [74, 93]. As a result these mice become highly immuno-deficient. Due to their inability to generate GCs, alymphoplastic mice such as LTβR$^{-/-}$, LTα$^{-/-}$, NIK$^{aly/aly}$ or NIK$^{-/-}$ are all defective in immunoglobulin (Ig) class switch and hypermutation [94-97].

One of the most widely used mouse strains for studying the role of SLTs during immune responses is the *alymphoplasia* (NIK$^{aly/aly}$) strain, which carries a NIK point mutation. The mutation in NIK$^{aly/aly}$ mice was originally found as a spontaneous autosomal recessive mutation that leads to the absence of all LNs and PPs. NIK$^{aly/aly}$ spleens are devoid of well defined lymphoid follicles and the white pulp is atrophic [94]. The NIK mutation in NIK$^{aly/aly}$ mice lies in the C-terminal region (G855R substitution [98]), which is responsible for physical interaction with the upstream TRAFs and IKKα [99]. Hence, the NIK$^{aly/aly}$ kinase domain is still intact and capable to induce NFκB activation, but due to lacking upstream signaling the levels of nuclear p52 in several tissues and cell types are virtually ablated [59].

NIK$^{aly/aly}$ mice display impaired antibody responses and loss of CMI, demonstrated by their inability to reject allogeneic grafts or tumors [99-101]. IgM levels in NIK$^{aly/aly}$ mice are one-third of these found in heterozygote animals, and IgG and IgA are virtually absent [94]. NIK$^{aly/aly}$ T cells have been shown to be defective in secretion of Il-2 and GM-CSF [70]. They are limited in proliferation and Th17 differentiation and fail to become pathogenic in EAE or in transplantation models [70, 102-107].

Summarized, NIK$^{aly/aly}$ mice show several deficiencies in humoral and cell-mediated immune responses. These deficiencies have been largely explained by their lack of SLTs. However, the fact that NFκB-signaling plays a role in virtually all aspects of immunity presents another explanation for the immunodeficiency in NIK$^{aly/aly}$ mice that have partly disrupted NFκB-signaling. This possibility was strongly ignored in previous research. We observed that NIK$^{aly/aly}$ mice are resistant to the induction of T cell-mediated EAE. That finding renders the LN-less NIK$^{aly/aly}$ mice a very interesting tool, in which we disconnected the two pathways of disrupted SLTs and lesioned NIK-signaling, in order to study on one hand the requirements of SLTs and on the other hand the requirement of NIK-signaling in CMI.

SPECIFIC AIMS

For a long time, the immunodeficiency of NIK$^{aly/aly}$ mice was attributed to their lack of LNs and absent organization in spleen, as these structures are believed to be absolutely essential for the induction of adaptive immunity. However, the possibility that the immunodeficiency of NIK$^{aly/aly}$ mice is caused by the intrinsic lesion of NIK has been neglected.

In this thesis, I therefore aimed to dissect the cause of NIK$^{aly/aly}$ immunodeficiency by defining the following objectives:

(a) Assess the structural requirements for efficient humoral and cell-mediated immunity and the development of autoimmunity.

(b) Define the role of NIK-signaling in cell-mediated immunity.

RESULTS

NEO-LYMPHOID AGGREGATES IN THE ADULT LIVER CAN INITIATE POTENT CELL-MEDIATED IMMUNITY

Greter M[1]*, Hofmann J[1]*, Becher B[1]
PLoS Biol. 2009 May 5;7(5)

[1] *Institute of Experimental Immunology, University Zürich, Switzerland*
* These authors contributed equally to this work.

In this study, we describe that the immunodeficiency of NIK$^{aly/aly}$ mice is not due to the absence of SLTs, but due to the impact of the underlying genetic defect on cellular immunity. Using different strains of alymphoplastic mice and T-cell migration mutants, in an experimental paradigm in which the site of Ag-delivery is distant from the site of priming and again distant from the site of inflammation, we can detect both Th cell-driven autoimmune disease as well as systemic CTL-mediated anti-tumor immunity initiated through classical subcutaneous (s.c.) immunization/vaccination independent of SLTs. APCs present at the site of immunization migrate to and select the liver as a natural extra-lymphoid tissue for the initiation of CMI, which we propose to be an evolutionary hard-wired pathway already found in cold-blooded vertebrates. This, to this day undescribed alternative pathway, can potently drive CMI but fails to elicit B cell immunity, indicating that the immunization-induced T cell accumulation within conventional lymphoid organs mainly serves humoral immunity but that CMI can be initiated elsewhere.

AUTOIMMUNITY CANNOT BE INITIATED IN NIK$^{ALY/ALY}$ MICE

We first sought to determine whether LNs are an absolute requirement for the induction of a complex Th cell-driven autoimmune response initiated by the s.c. delivery of auto-antigen (Ag). EAE is a B cell-independent, T$_H$ cell-mediated demyelinating autoimmune disease of the CNS and serves as the animal model for MS. The conversion of T$_H$ cells from the naïve to effector state is vitally dependent on the structures provided by LNs [108, 109]. Cervical LNs are widely held to constitute the predominant intrinsic priming site for encephalitogenic T cells, based on the observation that these LNs support the expansion of PLP-TcR Tg T cells [110, 111]. However,

RESULT

draining inguinal LNs drive the polyclonal, endogenous T cell population after s.c. immunization with encephalitogenic peptides. To assess the role of SLTs in the transition of T_H cells from a naïve to effector state (T cell priming), we induced EAE in NIK$^{aly/aly}$ or NIK$^{aly/+}$ mice by s.c. immunization with MOG peptide and complete Freund adjuvant (MOG$_{35-55}$/CFA). Fig. 7A shows that NIK$^{aly/aly}$ mice are completely resistant to EAE compared to NIK$^{aly/+}$ control mice (the latter developing normal SLTs as NIK is haplo-sufficient). To verify the notion that pathogenic T cells cannot be raised in NIK$^{aly/aly}$ mice, they were immunized s.c. with MOG$_{35-55}$. 11 dpi, splenocytes were harvested, MOG$_{35-55}$-reactive cells were expanded *in vitro* and subsequently transferred into NIK$^{aly/aly}$ as well as NIK$^{aly/+}$ recipients. Fig. 7B shows that only cells derived from NIK$^{aly/+}$ donors were able to induce disease regardless whether the recipients had SLTs (NIK$^{aly/+}$) or not (NIK$^{aly/aly}$). In contrast, MOG$_{35-55}$-reactive T cells derived from NIK$^{aly/aly}$ donors were not pathogenic and did not mediate CNS-inflammation.

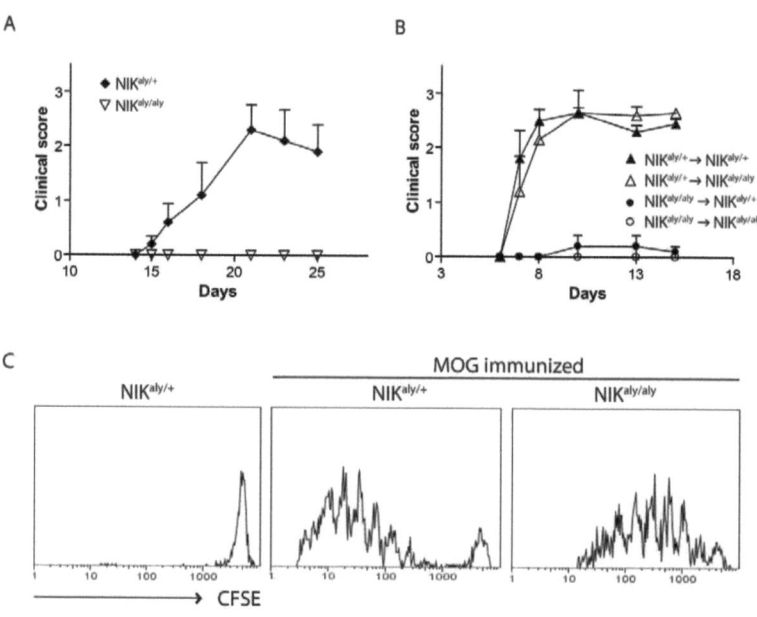

Figure 7. NIK$^{aly/aly}$ mice are resistant to the development of EAE. (A) EAE was induced by active immunization with MOG$_{35-55}$/CFA in NIK$^{aly/aly}$ (▽) and NIK$^{aly/+}$ mice (◆). **(B)** EAE was induced by adoptive transfer of pathogenic T cells derived from MOG-immunized NIK$^{aly/aly}$ or NIK$^{aly/+}$ donors into NIK$^{aly/aly}$ or NIK$^{aly/+}$ recipients. NIK$^{aly/+}$ into NIK$^{aly/+}$: ▲, NIK$^{aly/+}$ into NIK$^{aly/aly}$: △, NIK$^{aly/aly}$ into NIK$^{aly/+}$: ●, NIK$^{aly/aly}$ into NIK$^{aly/aly}$: ○. Shown is a representative of two individual experiments (n ≥ 5 mice/group). **(C)** NIK$^{aly/+}$ and NIK$^{aly/aly}$ mice were injected with 20x10^6 CFSE-labeled splenocytes i.v. derived from 2D2 transgenic mice and immunized s.c. with MOG$_{35-55}$/CFA. 4 dpi splenocytes were analyzed by flow cytometry by gating on 2D2$^+$ cells. Results are representative of 2 individual experiments (n = 2 mice/group).

To assess the capacity of LN-less mice to initiate T cell expansion in response to s.c. delivered Ag, CFSE-labeled TCR transgenic T cells (2D2) specific for the encephalitogenic MOG$_{35-55}$ peptide [112] were adoptively transferred into either NIK$^{aly/aly}$ or NIK$^{aly/+}$ mice prior to immunization with their cognate Ag. After 4 days, splenocytes were analyzed for T cell expansion by flow cytometry (Fig. 7C). Ag-specific T cell proliferation can be observed in NIK$^{aly/aly}$ mice, however, they display slightly delayed kinetics in comparison to NIK$^{aly/+}$ mice. Similar results were obtained with Ovalbumin (OVA) TCR transgenic T cells (OTII) transferred into NIK$^{aly/aly}$ and NIK$^{aly/+}$ mice (not shown), indicating that T cell expansion can be initiated independent of SLTs, while efficient effector function is dependent on the microenvironment provided by SLTs.

INDUCTION OF PRODUCTIVE T CELL IMMUNITY IN THE ABSENCE OF SLTS

The fact that NIK$^{aly/aly}$ mice do not develop T cell-driven autoimmune disease could be explained by their inability to prime self-reactive T cells i) due to the lack of dedicated draining LNs [108, 109], or ii) due to a direct impact of the NIK mutation on immune cells [105, 106]. In order to define whether their EAE-resistance is due to the lack of LNs or an intrinsic defect of NIK$^{aly/aly}$ mice to prime T cells, we generated a series of bone marrow (BM)-chimeric mice. To restrict the NIK mutation to the hematopoietic system, lethally irradiated NIK$^{aly/+}$ mice were injected with BM cells from NIK$^{aly/aly}$ donor mice (NIK$^{aly/aly}$→NIK$^{aly/+}$). Conversely, to conserve the developmental structural defects, without the NIK-lesion of the hematopoietic compartment, NIK$^{aly/aly}$ mice were reconstituted with BM cells of normal NIK$^{aly/+}$ donors (NIK$^{aly/+}$→NIK$^{aly/aly}$). As previously reported, spontaneous development of lymphoid tissues in NIK$^{aly/aly}$ recipients upon reconstitution was expectedly not detected [113].

Surprisingly, we discovered that NIK$^{aly/+}$→NIK$^{aly/aly}$ BM-chimeras were fully susceptible to EAE after s.c. immunization with MOG$_{35-55}$ (Fig. 8A), clearly demonstrating that s.c. immunization can mount a productive T cell-driven autoimmune response even in the absence of draining LNs. Using the reciprocal approach, by generating NIK$^{aly/+}$→NIK$^{aly/+}$ (wt-NIK immune system and normal SLTs) as well as NIK$^{aly/aly}$→NIK$^{aly/+}$ BM-chimeras (NIK-deficient immune system and normal SLTs), we found that the NIK mutation lead to EAE resistance, even when the lymphoreticular compartment is unperturbed (Fig. 8B). This finding clearly demonstrates that the reported immunodeficiency of NIK$^{aly/aly}$ mice can largely be explained by the requirement of NIK for the initiation of immunity rather than the lack of LNs. In support of this, we found that unmanipulated LTβR$^{-/-}$ mice, which also lack all LNs but have normal NIK function, are also fully susceptible to EAE (Fig. 8C).

RESULT

The formation of IFNγ and IL-17-secreting auto-reactive T cells has been demonstrated to be a prerequisite for the development of autoimmunity [11]. In NIK$^{aly/aly}$→NIK$^{aly/+}$ mice we observed a substantial reduction in IL-17 and IFNγ producing cells compared to the control mice NIK$^{aly/+}$→NIK$^{aly/+}$ (Fig. 8D) indicating that the resistance to EAE in the absence of NIK could be related to the function of NIK in T cell polarization. The mechanistic underpinnings of this phenomenon are currently being investigated, but it is clear that the loss of NIK signaling impairs the capacity of NIK$^{aly/aly}$ mice to generate pathogenic T$_H$ cells regardless of their structural defects.

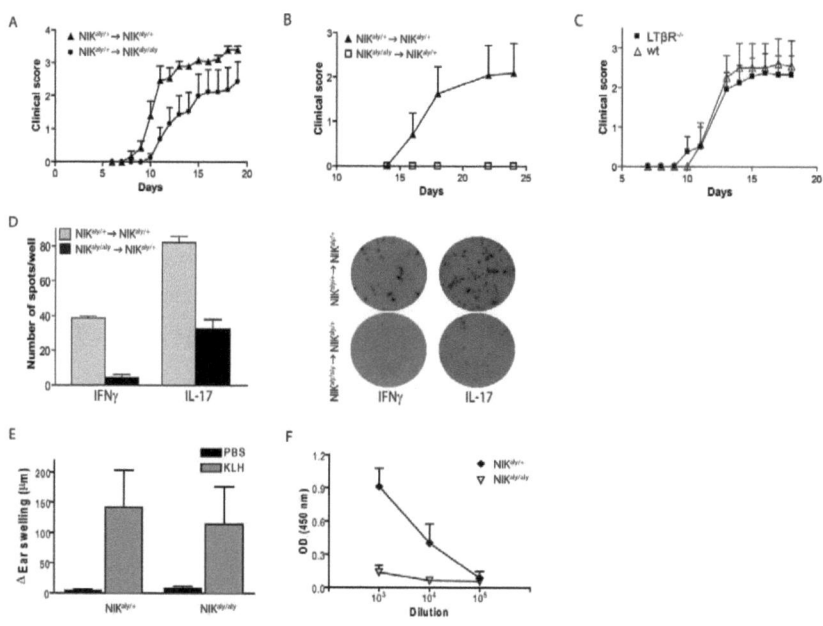

Figure 8. SLTs are crucial for B but not T cell mediated immune responses. (A-B) EAE progression in BM-chimeras immunized s.c. with MOG$_{35-55}$/CFA. (A) NIK$^{aly/+}$ → NIK$^{aly/+}$: ▲, NIK$^{aly/+}$ → NIK$^{aly/aly}$: ●. (B) NIK$^{aly/aly}$ → NIK$^{aly/+}$: □, NIK$^{aly/+}$ → NIK$^{aly/+}$: ▲. (C) EAE was induced by active immunization with MOG$_{35-55}$/CFA of LTβR$^{-/-}$ mice (■) and wt mice (Δ). Shown are representatives of three individual experiments (n≥5/group) ±SEM. (D) LN-derived cells were obtained from NIK$^{aly/aly}$ → NIK$^{aly/+}$ (black bars) and NIK$^{aly/+}$ → NIK$^{aly/+}$ (grey bars) BM-chimeras 21 dpi with MOG$_{35-55}$/CFA and rechallenged *in vitro* with 50 μg/ml MOG$_{35-55}$ peptide to reveal IFNγ and IL-17 secreting cells using Elispot. Shown is a representative of 2 individual experiments (n=3 /group ± SEM). (E) DTH responses were induced by s.c. immunization with KLH/CFA of NIK$^{aly/aly}$ and NIK$^{aly/+}$ mice. 11dpi, the mice were challenged by intradermal injection of KLH (grey bars), or PBS (black bars) into the ear. Swelling was measured 24h post challenge using a precision calliper and shown is the increase of ear swelling over baseline of a representative of 3 independent experiments (n ≥ 2 /exp). (F) Sera was collected from KLH-immunized NIK$^{aly/aly}$ (∇) and NIK$^{aly/+}$ mice (♦) mice on 12 dpi and analyzed for the presence of total anti-KLH Abs by ELISA. Results are representative of 3 independent experiments (*n* ≥ 2 mice/group).

B AND T CELLS HAVE DIFFERENT STRUCTURAL REQUIREMENTS FOR PRIMING AND MATURATION

Given the dogma that in mammals, CMI initiated by s.c. or intramuscular Ag-delivery requires the presence of SLTs, it is feasible that the remaining SLT (i.e. the spleen) in NIK$^{aly/+}$→NIK$^{aly/aly}$ BM-chimeras compensates for the absence of LNs. In order to test this notion, we splenectomized NIK$^{aly/+}$→NIK$^{aly/aly}$ BM-chimeras (NIK$^{aly/+}$→NIK$^{aly/aly\ spl}$) 14 days prior to the induction of EAE. Upon immunization, NIK$^{aly/+}$→NIK$^{aly/aly\ spl}$ mice developed EAE with the same disease severity as control mice (Table 1). We noted a slight delay in disease onset when all SLTs are absent, while histopathological analysis of diseased mice revealed no difference between NIK$^{aly/+}$→NIK$^{aly/+}$ and NIK$^{aly/+}$→NIK$^{aly/aly\ spl}$ mice (Fig. 9).

Table 1. Mice devoid of SLTs are fully susceptible to EAE.

BM-Chimeras	Incidence (%)	Mean Day of Disease Onset[a]	Mean Maximal Clinical Score[a]
aly/+→aly/+	94.7 (18 of 19)	11.1 (±0.4)	3.0 (±0.1)
aly/+→aly/+^{spl}	75 (9 of 12)	11.8 (±0.2)	3.0 (±0.3)
aly/+→aly/aly	65.2 (15 of 23)	13.7 (±0.7)	3.3 (±0)
aly/+→aly/aly^{spl}	64.7 (11 of 17)	15.8 (±0.8)	3.4 (±0.1)
aly/aly→aly/+	0 (0 of 10)	—	—

[a] of diseased animals (±SEM).

Figure 9. Inflammatory lesions in the CNS of mice lacking SLTs. H&E stainings of spinal cord sections of diseased NIK$^{aly/+}$ → NIK$^{aly/+}$ and NIK$^{aly/+}$ → NIK$^{aly/aly\ spl}$ BM-chimeras. Lower row represents higher magnification of the insert in upper row. Bar in upper row: 200 μm and in lower row: 50 μm.

In contrast to T cell activation, we found that B cell activation requires the structural environment provided by SLTs. To investigate the impact of immunization on T vs. B cell responses, we used Keyhole limpet hemocyanin (KLH) as a model foreign Ag to elicit delayed-type hypersensitivity (DTH) responses. NIK$^{aly/aly}$ as well as NIK$^{aly/+}$ mice were immunized with KLH and 11 dpi, they were challenged by intradermal injection with KLH into the ear. As illustrated in Fig. 8E, both groups were able to mount a solid DTH reaction measured by ear swelling, which was only marginally lower in NIK$^{aly/aly}$ than in NIK$^{aly/+}$ mice. However, in contrast to ear-swelling, which is indicative of CMI, NIK$^{aly/aly}$ mice did not mount Abs against KLH when compared to NIK$^{aly/+}$ mice demonstrating that the development of a humoral immune response is ablated in the absence of lymphoreticular structures (Fig. 8F). We could reproduce functional DTH responses using other Ags including Ovalbumin and myelin oligodendrocyte glycoprotein peptide (MOG$_{35-55}$) (not shown). Similarly in our EAE paradigm using BM-chimeras, while control mice (NIK$^{aly/+}$→NIK$^{aly/+}$ and NIK$^{aly/+}$→NIK$^{aly/+\ spl}$) elicit high Ab titers, anti-MOG Abs are virtually absent in mice without LNs (either NIK$^{aly/+}$→NIK$^{aly/aly}$ or NIK$^{aly/+}$→NIK$^{aly/aly\ spl}$) (Fig. 10A). Analysis of isotype subtypes revealed that in splenectomized alymphoplastic mice, elevated anti-MOG IgM could be detected which has previously been reported [114-116], while class switching to IgG could not be observed (Fig. 10B).

Taken together, and in agreement with the notion that SLTs are vital for B cell activation, for the generation of high-affinity Igs and class switching, highly organized SLTs are obligatory, while potent cellular immunity can be induced successfully upon s.c. immunization even in the absence of SLTs.

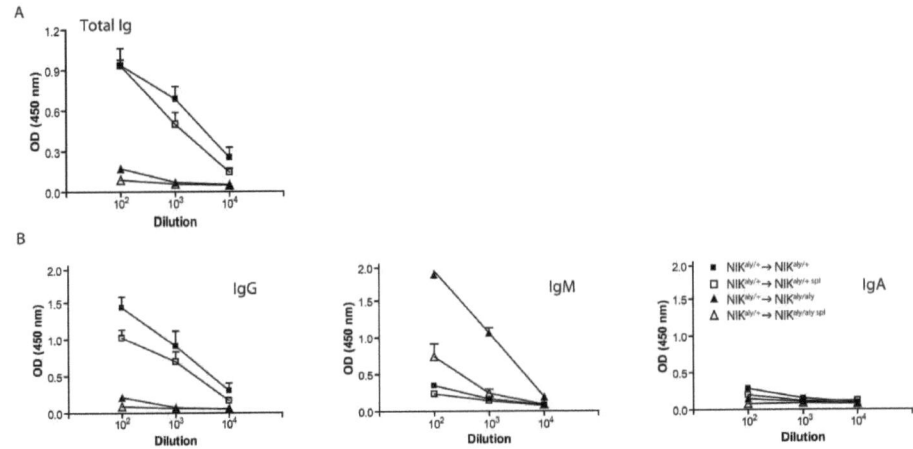

Figure 10. Ab-response to s.c. auto-Ag depends on the presence of dedicated lymphoid structures. (A-B) Titers of anti-MOG Abs (total Ig, IgG, IgM and IgA) determined from sera of diseased BM-chimeras immunized s.c. with MOG_{35-55}/CFA by ELISA. $NIK^{aly/+} \rightarrow NIK^{aly/+}$: ■, $NIK^{aly/+} \rightarrow NIK^{aly/+\ spl.}$: □, $NIK^{aly/+} \rightarrow NIK^{aly/aly}$: ▲, $NIK^{aly/+} \rightarrow NIK^{aly/aly\ spl.}$: △. (E) Total Ig, (F) IgG, IgM, IgA. Shown is a representative of 3 individual experiments (n=3 /group ± SEM).

IN THE ABSENCE OF SLTs, SUBCUTANEOUSLY DELIVERED AG IS TRANSPORTED INTO THE LIVER

Since the loss of SLTs in $NIK^{aly/aly}$ BM-chimeric mice does not hinder the development of T cell immunity, we wanted to determine at which alternative site T cell priming could take place and to which organ the Ag travels from the site of immunization (s.c.). Therefore, $NIK^{aly/aly}$ BM-chimeras were injected s.c. with Yellow Green (YG) carboxylate microspheres emulsified in CFA. 7 dpi various organs were isolated and analyzed for the presence of fluorescent cells by flow cytometry. Fig. 4A shows that in control mice ($NIK^{aly/+} \rightarrow NIK^{aly/+}$) fluorescently labeled APCs were exclusively detected in LNs upon s.c. immunization. It was previously shown that the BM has the capacity to drive an enriched population of high affinity TCR Tg T cells in response to blood-borne Ag [117]. As expected, upon i.v. delivery of Ag, the vast majority of it accumulates in the spleen, BM and liver, regardless of the presence of SLTs (Fig. 11). However, after (s.c.) immunization of $NIK^{aly/+} \rightarrow NIK^{aly/aly\ spl}$ BM-chimeras lacking SLTs, APCs carrying fluorescent microspheres migrate primarily to the liver and not the other organs analyzed (thymus, CNS & gut are not shown) (Fig. 12A). Only a small amount of Ag reaches the liver when draining SLTs are present. Next, we wanted to determine the means of the Ag transport from the s.c. reservoir to the liver. To determine whether the Ag diffuses to the liver or is actively transported by APCs, $NIK^{aly/+} \rightarrow NIK^{aly/+}$ and $NIK^{aly/+} \rightarrow NIK^{aly/aly\ spl}$ chimeric mice were separated into two groups. One received YG micropsheres/CFA in the left flank and polychromatic red (PR) microspheres /CFA in the right flank. The other group received a mixture of YG and PR coupled beads in both flanks (see scheme in Fig. 12B). After 7 days, mice were sacrificed, perfused and a single cell suspension of livers, LNs and spleens was generated for cytofluorometric analysis. We found that the mixture of PR/YG-coupled beads generated a large proportion of dual-labeled CD11b as well as CD11c positive APCs. Conversely, the injection of either PR or YG-coupled microspheres into each flank revealed merely single-labeled APCs in the liver. The presence of single-labeled cells within the liver strongly suggests that the Ag is delivered to the liver by the migration of APCs initially present at the site of immunization. Passive diffusion of the Ag from the site of immunization via the blood-stream to the liver cannot be fully excluded, but is evidently not the dominating means of Ag delivery. In addition, only a negligible amount of Ag reaches the liver when dedicated SLTs are present (Fig. 12). We could also confirm these finding by using soluble FITC painted on shaved flanks (without the adjuvant CFA). 24h after FITC$^+$ skin painting, we found FITC$^+$ APCs primarily in the liver again supporting the notion that the liver can serve as an alternative Ag-presenting site when draining LNs are not available (Fig. 12C).

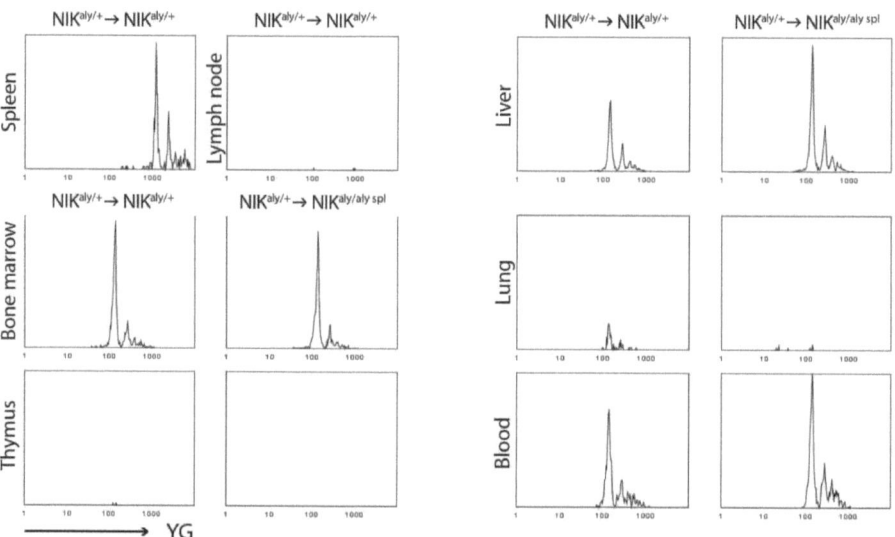

Figure 11. Intravenously delivered Ag accumulates in the spleen, BM and liver. NIK^{aly} BM-chimeras were injected i.v. with YG microspheres and various organs were analyzed by FACS for the presence of fluorescently labeled APCs 7 dpi. Data represent one of 3 individual experiments.

RESULT

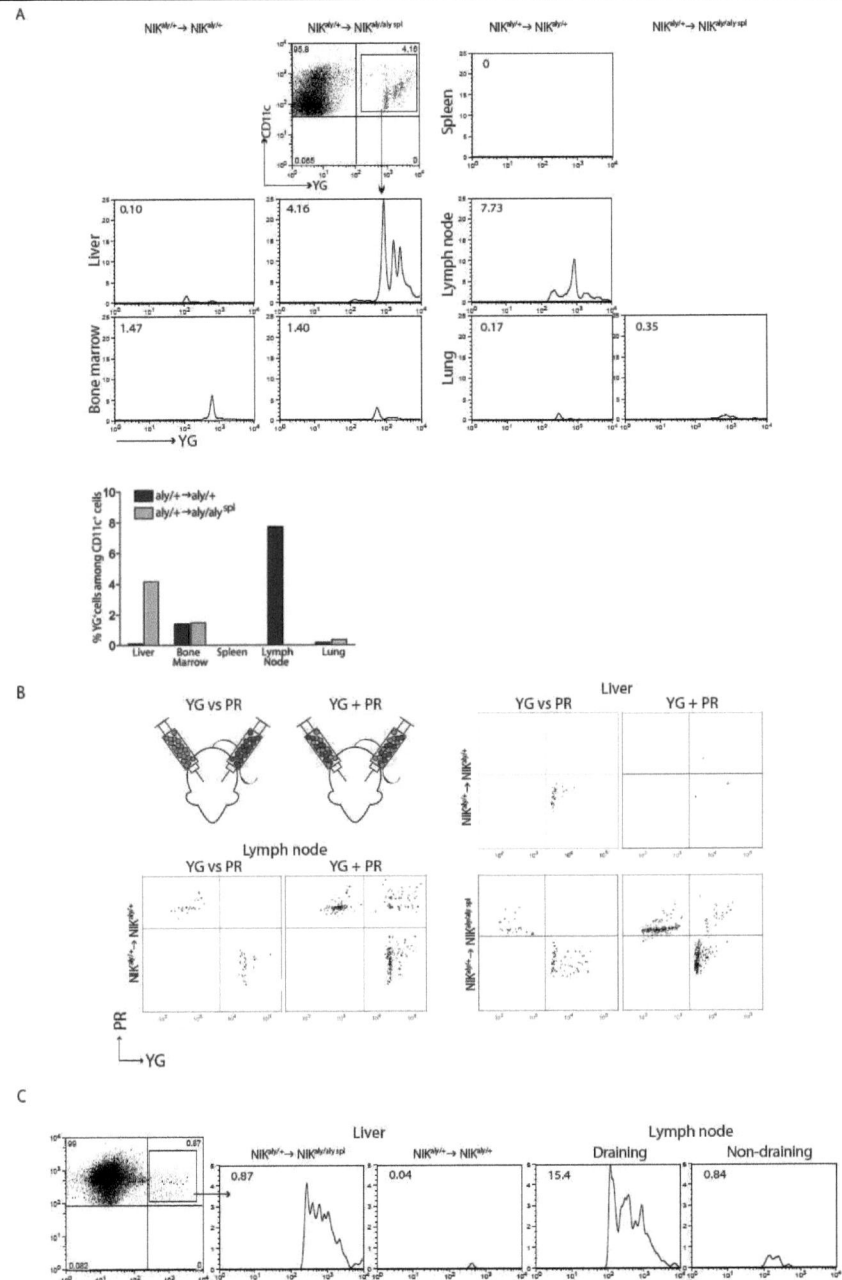

Figure 12. Ag-laden APCs migrate to the liver in the absence of SLTs. (A) NIK^{aly} BM-chimeras were injected s.c. with YG microspheres/CFA and various organs were analyzed by FACS for the presence of fluorescently labeled APCs 7 dpi. Data represent one of 3 individual experiments. **(B)** NIK^{aly} BM-chimeras were injected s.c. with either a mixture of YG and PR beads (YG+PR) into both flanks or YG beads into one flank and PR beads into the other flank (YG vs PR). 7 dpi, livers and LNs (only in NIK$^{aly/+}$ → NIK$^{aly/+}$ mice) were analyzed by FACS for single (YG or PR) or double (YG and PR) positive APCs (gated on CD11c$^+$ and CD11b$^+$ cells). **(C)** NIK^{aly} BM-chimeras were painted on the shaved flanks with 100 µL of 4 mg/mL FITC dissolved in 1:1 acetone:dibutylphalate. After 24h, livers and in NIK$^{aly/+}$ → NIK$^{aly/+}$ mice draining and non-draining inguinal LNs were analyzed by FACS for the presence of FITC$^+$ cells.

EXTRA-LYMPHOID AGGREGATES IN THE LIVER HOST T CELL/APC ENCOUNTERS

In order to determine whether lymphoid-like structures can be found in the liver, we analyzed the livers of immunized mice by histology (d7). Livers of NIK$^{aly/+}$→NIK$^{aly/aly\ spl}$ BM-chimeras showed massive infiltration of leukocytes in comparison to NIK$^{aly/+}$→NIK$^{aly/+}$ control mice (Fig. 13). Histological analysis displays dendritic cells (DCs) in close proximity to T cells in the infiltrated periportal areas of the liver indicative of T cell priming by Ag-laden APCs (Fig. 13B). In spite of the stromas inability to respond to LTα/β, detailed histological analysis revealed the presence of VCAM and ICAM in the infiltrates as well as B cells (Fig. 14) and even the presence of CXCL13 transcripts indicative of aggregates ability to recruit B cells (not shown). However, no evidence for GC formation could be obtained (Fig. 14).

We also transferred TcR Tg T cells from Luciferase-2D2 (Luc-2D2) mice into recipient BM-chimeras and observed the accumulation of Ag-responsive T cells in the liver 2 dpi with MOG$_{35-55}$/CFA by bioluminescence imaging (Fig. 15A). Fig. 15B shows that the number of DCs (CD11c$^+$) and adoptively transferred 2D2 T cells (CD4$^+$/Vβ11$^+$) is drastically increased in the liver in mice lacking SLTs.

In order to demonstrate that the observed lymphocyte accumulations in the liver can support cell-expansion, we injected naïve (CD62L$^+$) CD4$^+$ T cells derived from 2D2 Tg mice into NIK$^{aly/aly}$ BM-chimeras and subsequently immunized them with MOG$_{35-55}$/CFA. 5 dpi, livers were analyzed for Ag-specific CD4$^+$ T cell proliferation. Even in normal mice, we find a large number of expanded T cells within the liver (Fig 15C), but one could argue that they have immigrated from their initial priming site, the draining LN. However, in the absence of SLTs, the livers of NIK$^{aly/+}$→NIK$^{aly/aly\ spl}$ BM-chimeric mice are sufficient to propagate Ag-driven T cell expansion and accumulation. In order to confirm that Ag-specific T cell proliferation occurs *in situ* in the liver, we administered BrdU i.p. into NIK$^{aly/aly}$ BM-chimeras 7 dpi with MOG$_{35-55}$/CFA. 30 min after BrdU injections, the mice were sacrificed and livers were analyzed for proliferating (BrdU$^+$) CD4$^+$ T cells by flow cytometry. Fig. 15D and E reveal the presence of BrdU$^+$ cells in the livers of both NIK$^{aly/+}$→NIK$^{aly/+}$ and NIK$^{aly/+}$→NIK$^{aly/aly\ spl}$ BM-chimeras. The number of BrdU$^+$ T cells in the liver is increased in NIK$^{aly/+}$→NIK$^{aly/aly\ spl}$ BM-chimeras compared to the controls. The fact that we found such a rapid (30 min) emergence of proliferating T cells even in normal mice in which SLTs are present, indicates that some degree of liver-initiated CMI occurs simultaneously to the priming within draining LNs .

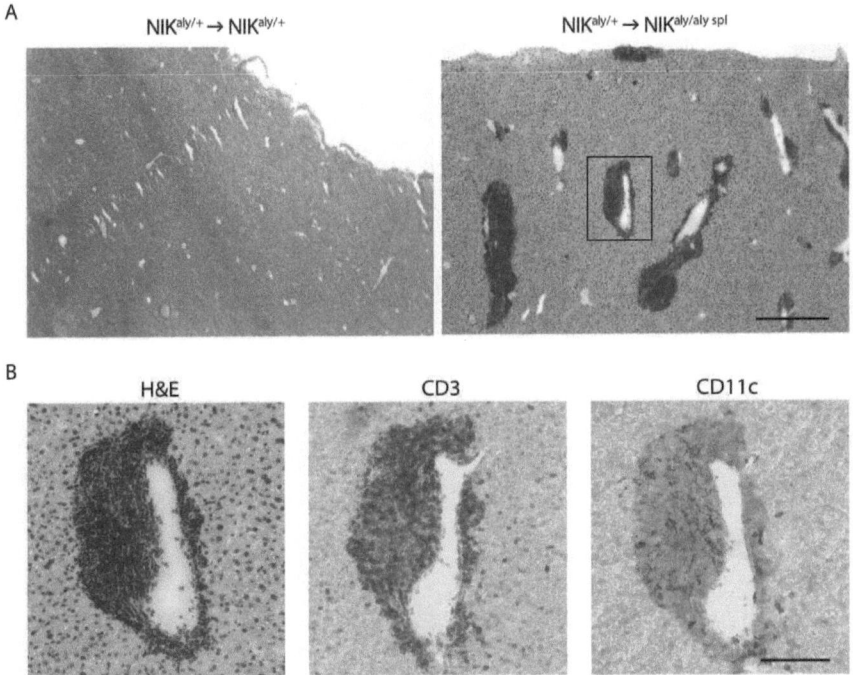

Figure 13. Extra-lymphoid aggregates in the liver host T cells and APCs. (A) Liver cryosections from NIK^{aly} BM-chimeras immunized s.c. with MOG$_{35-55}$ (d7) were stained with H&E. Bar: 500 μm. **(B)** Higher magnification image indicated by the region square in (A) stained with H&E and mAbs against CD3 and CD11c. Bar: 100 μm.

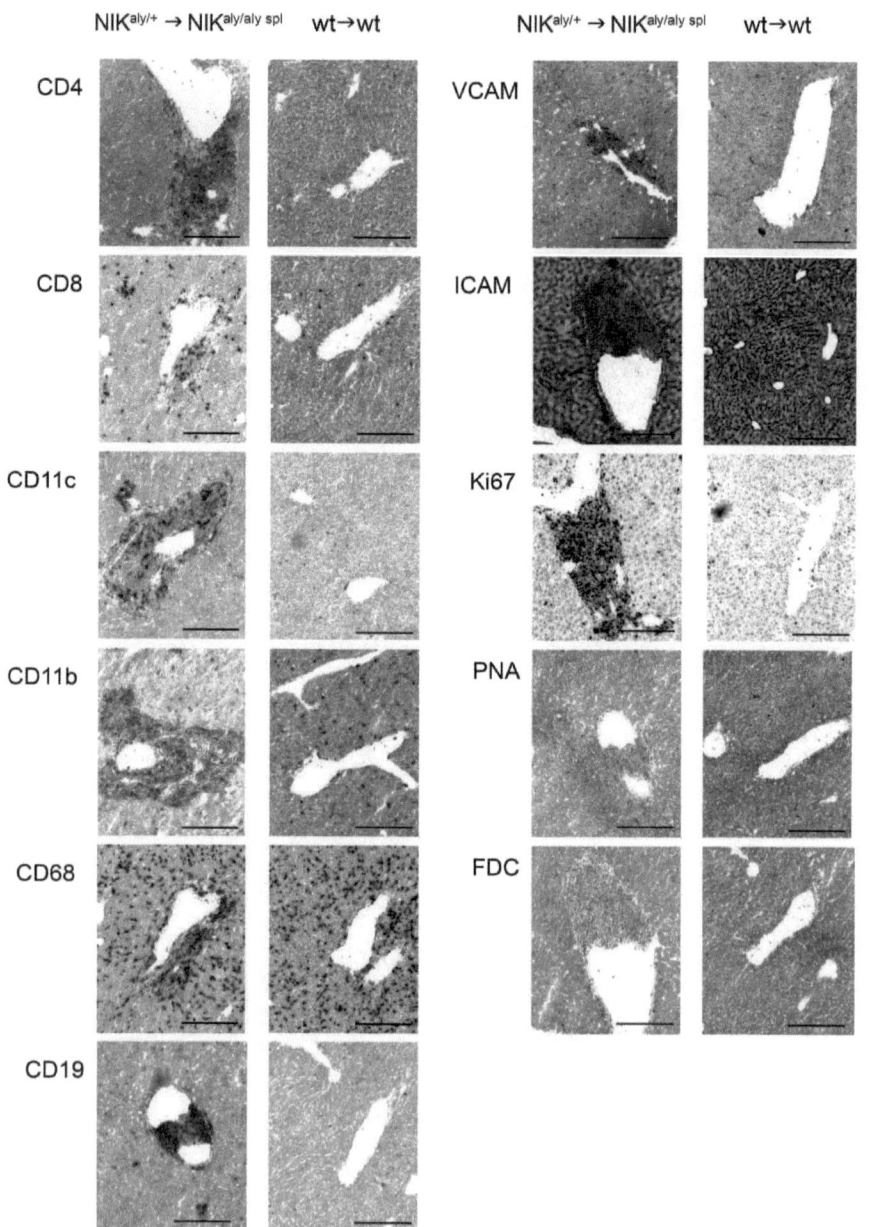

Figure 14. Expression of lymphoid structure markers in livers of NIK^{aly} BM-chimeric mice. Liver cryosections from NIK^{aly} BM-chimeras immunized s.c. with MOG$_{35-55}$ (d11) were stained with antibodies against CD4, CD8, CD11b, CD11c, CD19, CD62L, CD68, FDC, ICAM, Ki67, PNA and VCAM. Bar: 200 μm

RESULT

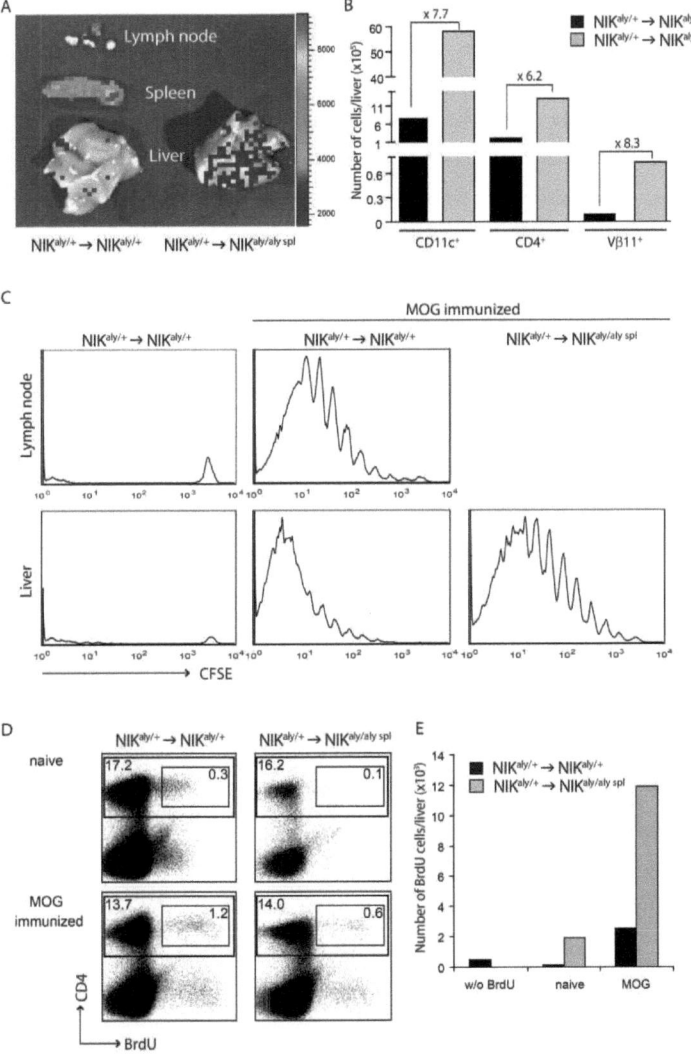

Figure 15. (A-B) Accumulation and Ag-specific T cell expansion in the liver. NIK^{aly} BM-chimeras were injected i.v. with 8x10^6 Luc-2D2 Tg CD4$^+$ T cells and immunized s.c. with MOG$_{35-55}$/CFA. **(A)** 2 dpi, mice were injected with luciferin and after 10 minutes sacrificed. Livers and in control mice LNs and spleen were isolated and images were acquired by bioluminescence imaging to reveal the accumulation of the injected luciferase positive (Luc-2D2) cells. **(B)** Absolute numbers of liver-invading DCs and Ag-specific T cells assessed from the percentage of CD11c$^+$, CD4$^+$ and Vβ11$^+$ cells analyzed by flow cytometry. Numbers above the graph indicate the fold-increase of liver-invading cells of NIK$^{aly/+}$ → NIK$^{aly/aly\ spl}$ (grey) over NIK$^{aly/+}$ → NIK$^{aly/+}$ (black). **(C)** NIK$^{aly/+}$ → NIK$^{aly/+}$ and NIK$^{aly/+}$ → NIK$^{aly/aly\ spl}$ BM-chimeras were injected with 8x10^6 CFSE-labeled naïve (CD62L$^+$) CD4$^+$ T cells derived from 2D2 transgenic mice and immunized s.c. with MOG$_{35-55}$/CFA. 5 dpi LNs (only in NIK$^{aly/+}$ → NIK$^{aly/+}$) and liver-invading cells were analyzed by flow cytometry by gating on 2D2$^+$ cells. **(D-E)** NIK$^{aly/+}$ → NIK$^{aly/+}$ and NIK$^{aly/+}$ → NIK$^{aly/aly\ spl}$ BM-chimeras were immunized s.c. with MOG$_{35-55}$/CFA. **(D)** 7 dpi, BM-chimeras were injected with BrdU i.p. 30 min after BrdU injection, liver-invading cells were analyzed by flow cytometry for BrdU$^+$ CD4$^+$ cells. **(E)** Absolute numbers of liver-invading BrdU$^+$ CD4$^+$ T cells assessed by flow cytometry. NIK$^{aly/+}$ → NIK$^{aly/aly\ spl}$ (grey) and NIK$^{aly/+}$ → NIK$^{aly/+}$ (black).

The Adult Liver can Support T cell, But not B cell Priming

In contrast to our findings which show that mice lacking SLTs do not generate high-affinity Ab-responses, intranasal influenza infection of splenectomized $LT\alpha^{-/-}$ mice reconstituted with wt stem cells for instance can initiate the formation of extra-lymphoid follicles within the lung which support some degree of B cell maturation and Ab-secretion [24,26]. One possible explanation for these contrasting observations regarding Ab-production is that in our case, stroma cells such as FDCs cannot signal through $LT\beta R$ due to the mutation within NIK and that this could be the reason for our inability to observe GC formation and Ab-secretion, while Moyron-Quiroz et al. used mice in which the stroma compartment can be engaged by $LT\alpha/\beta$ [114, 118]. To definitively address whether the stoma's inability to signal through NIK is the reason for the weak B cell response, we obtained LN-deficient $LT\alpha^{-/-}$ mice and reconstituted their hematapoietic system with wt stem cells. The resulting chimeras were splenectomized and lacked all peripheral SLT (analogous to the $NIK^{aly/+} \rightarrow NIK^{aly/aly \ spl}$). Yet in contrast to $NIK^{aly/+} \rightarrow NIK^{aly/aly \ spl}$, wt$\rightarrow LT\alpha^{-/-spl}$ chimeras have normal stromal cell function and FDCs are capable of responding to $LT\alpha/\beta$. These mice were immunized s.c. and the formation of B cell maturation and Ab production was analyzed. Fig. 16A demonstrates that these wt$\rightarrow LT\alpha^{-/-spl}$ chimeras behave exactly like $NIK^{aly/+} \rightarrow NIK^{aly/aly \ spl}$ in regards to their inability to generate high Ab titers and to class switch.

In a comparative fashion, we analyzed the histological parameters of wt$\rightarrow$wt, $NIK^{aly/+} \rightarrow NIK^{aly/aly \ spl}$ and wt$\rightarrow LT\alpha^{-/-spl}$ chimeras (Fig. 16B). While only alymphoplastic mutants revealed the presence of lymphoid aggregates surrounding periportal areas of the liver, neither FDCs nor PNA positive clusters could be found, again supporting the notion that the surrogate structures in the liver support T cell function but fail to initiate the formation of GCs needed for Ab-affinity cell maturation and class switching. Lastly, the large number of $Ki67^+$ cells within the liver aggregates again support our conclusion, that active proliferation within the liver can be induced by s.c. immunization (Fig 16B).

RESULT

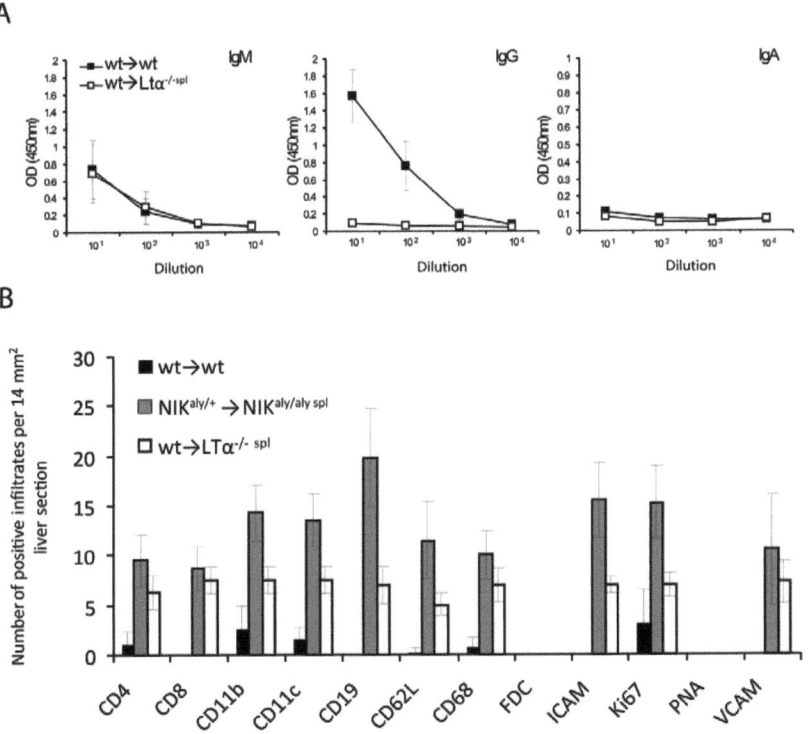

Figure 16. Surrogate liver aggregates support CMI, but not B cell maturation.
Wt→wt and wt→LTα$^{-/-spl}$ BM-chimeric mice were immunized s.c. with MOG$_{35-55}$/CFA. (A) 11dpi titers of anti-MOG Abs (IgG, IgM and IgA) were determined from sera by ELISA (n=4 mice/group). (B) Liver sections from wt→wt, wt→LTα$^{-/-spl}$ and NIK$^{aly/+}$ → NIK$^{aly/aly\;spl}$ BM-chimeras were stained with Abs against CD4, CD8, CD11b, CD11c, CD19, CD62L, CD68, FDC, ICAM, Ki67, PNA and VCAM. Positively stained infiltrated areas of 14mm^2 liver sections were counted (n=4 mice/group).

PRIMING OF CYTOTOXIC ANTITUMOR T CELLS IS INDEPENDENT OF SLTS

While we have demonstrated the development of T$_H$ cell-driven autoimmune disease in mice lacking SLTs, we wanted to elucidate whether these mice are also capable of inducing successful CTL immunity. We used the B16.F10 murine melanoma model, which represents a lethal and poorly immunogenic cancer. Irradiated GM-CSF expressing B16.F10 cells are used as s.c. vaccine to initiate potent CD8$^+$-antitumor immunity against live parental B16.F10 tumor cells [119]. We injected irradiated B16.F10-GM-CSF cells s.c. into one flank of NIK$^{aly/+}$ → NIK$^{aly/+}$ and NIK$^{aly/+}$ → NIK$^{aly/aly\;spl}$ chimeric mice. 12 dpi, mice were challenged with parental B16.F10 cells injected into the opposite flank. Fig. 17A shows that NIK$^{aly/+}$ → NIK$^{aly/aly\;spl}$ chimeric mice can elicit potent antitumor CTL responses revealed by the inhibition of tumor growth. Next we transferred CFSE-labeled MHC class I restricted OVA-TcR Tg OTI T cells into NIK$^{aly/+}$ → NIK$^{aly/aly\ spl}$ and NIK$^{aly/+}$ → NIK$^{aly/+}$ BM-chimeric mice and subsequently injected irradiated B16.F10 cells expressing OVA. 12 dpi, livers and in control animals also spleen and LNs were analyzed by FACS for Ag-specific CD8$^+$ T cell expansion. As demonstrated in Fig. 17B, proliferation of CD8$^+$ OTI cells was detected in the liver of mice lacking SLTs. Hence, even under conditions in which the draining LNs are considered a compulsory site hosting the encounter of captured Ag and infiltrating CD8$^+$ T cells, we can detect potent T cell responses, which originate in the liver when SLTs are absent.

LIVER FOLLICLES ARE INDUCED BY IMMUNIZATION AND ABERRANT HOMEOSTATIC T CELL MIGRATION

We next wanted to address the relevance of the liver to serve as an alternative priming site in a setting where LNs are present but T cell migration into LNs is defective. To this end, we analyzed *plt/plt* (paucity of LN T cells) mice, which display undisturbed B cell zones but severely abrogated T cell zones due to the loss of CCL19 and CCL21, which results in the inhibition of both naive T cell and DC homing into SLTs [120]. We found that p*lt/plt* mice too developed delayed but fulminant EAE after s.c. immunization with MOG$_{35-55}$/CFA (Fig. 17C). Examination of liver sections of immunized *plt/plt* mice again revealed lymphocyte aggregates consisting mainly of CD4$^+$ T cells and DCs within the liver (Fig. 17D).

RESULT

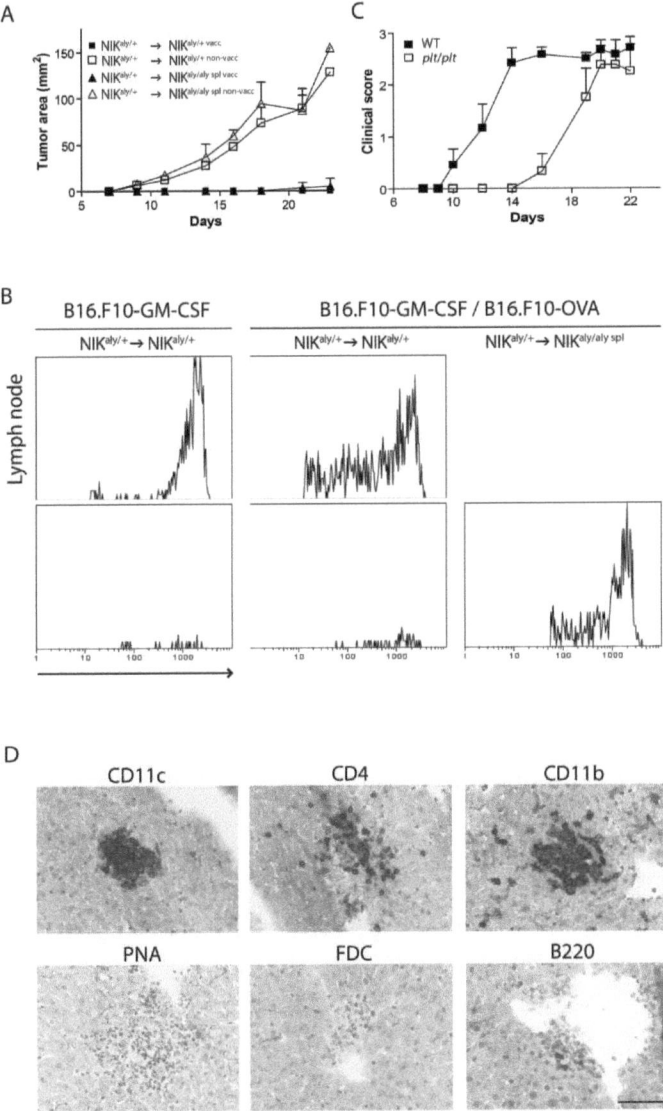

Figure 17. CD8⁺ T cell priming in the liver and lymphoid aggregates in *plt/plt* mice. (A) Tumor progression of NIK^{aly} BM-chimeras. Mice were vaccinated s.c. with 1x10⁶ irradiated GM-CSF-B16.F10 cells into one flank and 12d later, mice received 2x10⁵ live B16.F10-Luc cells into the opposite flank. Vaccinated NIK$^{aly/+}$ → NIK$^{aly/+}$: ■, non-vaccinated NIK$^{aly/+}$ → NIK$^{aly/+}$: □, vaccinated NIK$^{aly/+}$ → NIK$^{aly/aly\ spl}$: ▲, non-vaccinated NIK$^{aly/+}$ → NIK$^{aly/aly\ spl}$: △. (B) NIK$^{aly/+}$ → NIK$^{aly/+}$ and NIK$^{aly/+}$ → NIK$^{aly/aly\ spl}$ BM-chimeras were injected i.v. with 20x10⁶ CFSE-labeled splenocytes from OTI transgenic mice and s.c. injected with a mix of 1x10⁶ B16.F10-OVA and 1x10⁶ B16.F10-GM-CSF cells. 12 dpi LNs (only in NIK$^{aly/+}$ → NIK$^{aly/+}$) and liver-invading cells were analyzed by flow cytometry for the proliferation of CD8⁺ OTI cells (Vα2⁺). (C) EAE progression of *plt/plt* (□) and wt (■) mice immunized s.c. with MOG$_{35-55}$/CFA. (D) Liver cryosections from diseased *plt/plt* mice (C) were stained with mAbs against CD11c, CD11b, CD4, FDC, B220 and PNA. Bar: 100 μm.

55

Result

NIK Signalling in Dendritic Cells but not in T Cells is Required for the Development of Effector T Cells and Cell-Mediated Responses

Janin Hofmann[1], Florian Mair[1], Melanie Greter[1], Marc Schmidt-Supprian[2] and Burkhard Becher[1]
J Exp Med. 2011 Aug 29;208(9):1917-29. Epub 2011 Aug 1.

[1] *Institute of Experimental Immunology, University of Zurich, Winterthurerstrasse 190, 8057 Zurich, Switzerland*

[2] *Molecular Immunology and Signal Transduction, Max Planck Institute of Chemistry, Martinsried, Germany*

Even though the immunodeficiency in $NIK^{aly/aly}$ mice was thought to be due to their *alymphoplasia*, we have previously reported that the defect in CMI in $NIK^{aly/aly}$ mice is not connected to their lack of SLTs but that NIK activity is critical for cellular immune function [104]. CMI could be induced in the total absence of SLTs, when splenectomised $NIK^{aly/aly}$ mice were reconstituted with a wildtype (wt) hematopoietic system. However, CMI could not be induced in mice featuring normal SLTs but carrying the $NIK^{aly/aly}$ lesion only in hematopoietic cells. Although a critical function for NIK has been suggested specifically in T cells by several reports [70, 102, 103, 105, 106], the data presented here suggest that the loss of T cell function in $NIK^{aly/aly}$ mutants is not T cell intrinsic, but rather resulting from a defect in accessory leukocytes, namely DCs. We show here that NIK-lesioned T cells are indeed capable of secreting effector cytokines and acquiring full effector functions, depending on the thymic environment in which they develop. Instead of a T cell intrinsic lesion, loss of NIK in thymic DCs imprints a long-lasting halt in T cell effector function. If however, NIK-deficient T cells mature within a thymic environment which hosts NIK, T cells gain proper effector functions regardless of their NIK-deficiency. Thus, we propose that thymic DCs, which so far have only been implemented in negative selection [121] and the induction of natural regulatory T cells (nT_{regs}) [122], are capable of imprinting subsequent T cell effector function onto developing thymocytes. This ability of thymic DCs is strictly dependent on NIK.

Loss of NIK Function Results in Reduced T cell Proliferation, Differentiation and Production of Effector Cytokines.

We have recently reported that $NIK^{aly/aly}$ mice are resistant to the induction of EAE due to the loss of function of NIK within the hematopoietic compartment and more specifically due to a defect in T cell priming, independent of the lack of SLTs [104]. Furthermore, $NIK^{aly/aly}$ T cells have been reported to be defective in proliferation and secretion of Il-17, Il-2 and GM-CSF [70, 103, 105]. In line with these previous reports, we observed that *in vitro* polyclonally activated $NIK^{aly/aly}$ CD4$^+$ T cells produced less effector cytokines (Il-2, IFNγ, Il-17) whereas the production of Il-4 was unaffected (Fig. 18A). This suggests a T cell intrinsic impairment of T cell polarization conferred by the ablation of NIK-signaling. To further investigate the requirement of NIK for antigen-specific T cell activation, we crossed $NIK^{aly/aly}$ mice with TCR-transgenic 2d2 mice, in which the TCR recognizes the immuno-dominant epitope of the myelin oligodendrocyte glycoprotein (MOG$_{35-55}$). As expected, $NIK^{aly/aly}$-2d2 T cells also failed to secrete effector cytokines upon encountering their cognate antigen (Fig. 18B). To exclude the possibility that the observed defects were caused by the developmental malformations of SLTs in $NIK^{aly/aly}$ mice, we generated bone-marrow chimeric mice (BMCs) in which wt mice were reconstituted with hematopoietic stem cells from either $NIK^{aly/aly}$-2d2 or $NIK^{aly/+}$-2d2 mice. We found that even if most of the CD4$^+$ T cells carry the cognate antigen-specific TCR, $NIK^{aly/aly}$-2d2→wt BMCs retained their EAE-resistance upon MOG$_{35-55}$/CFA immunization (Fig. 18C), emphasizing the critical role of NIK-signaling for the development of autoimmune responses.

Since T cells from $NIK^{aly/aly}$-2d2→wt BMCs fail to acquire pathogenic properties, we addressed their behaviour in antigen-independent homeostatic expansion in lymphopenic $Rag1^{-/-}$ mice. After adoptive transfer of CD4$^+$ T cells from $NIK^{aly/aly}$-2d2→wt BMCs into $Rag1^{-/-}$ mice we observed a drastic reduction in homeostatic expansion when compared to T cells from $NIK^{aly/+}$-2d2→wt BMCs (Fig. 18D). Upon immunization of those mice with MOG$_{35-55}$/CFA, $NIK^{aly/aly}$-2d2 T cells further failed to respond to their cognate antigen, whereas control $NIK^{aly/+}$-2d2 T cells strongly expanded (Fig. 18E). In addition, $Rag1^{-/-}$ mice reconstituted with T cells from $NIK^{aly/aly}$-2d2→wt BMCs remained completely resistant to EAE, suggesting that $NIK^{aly/aly}$-2d2 T cells cannot be primed by NIK-sufficient accessory cells (Fig. 18F). Taken together, the data support the notion that NIK-deficiency indeed leads to a T cell intrinsic lesion.

RESULT

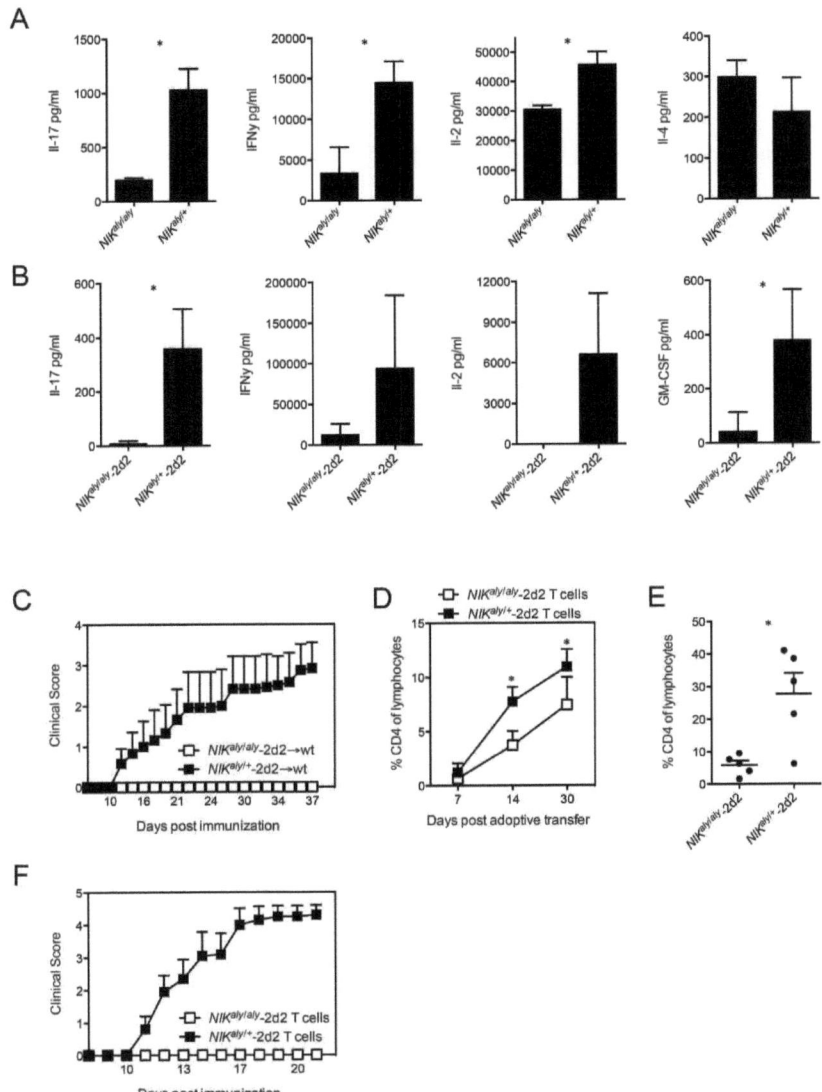

Figure 18: $NIK^{aly/aly}$ **mice are EAE resistant due to the lack of non-canonical NFκB-signaling in immune cells.** (A) CD4[+] T cells of $NIK^{aly/aly}$ or $NIK^{aly/+}$ mice were stimulated *in vitro* with plate-bound α-CD3/α-CD28 for 48 hours. Cytokine secretion was analyzed by ELISA. (B) Splenocytes of $NIK^{aly/aly}$-2d2 or $NIK^{aly/+}$-2d2 mice were stimulated *in vitro* with MOG_{35-55} and α-CD28 for 48 hours. Cytokine secretion was analyzed by ELISA. (C) $NIK^{aly/aly}$-2d2→wt or $NIK^{aly/+}$-2d2→wt BMC mice were immunized with MOG_{35-55}/CFA and monitored daily for clinical signs of EAE (n=6). (D-F) CD4[+] T cells of $NIK^{aly/aly}$-2d2→wt or $NIK^{aly/+}$-2d2→wt BMCs were transferred into $Rag1^{-/-}$ mice, respectively. Homeostatic expansion was observed by weekly FACS analysis of blood samples (D). 30 days after adoptive CD4[+] T cell transfer $Rag1^{-/-}$ mice were immunized with MOG_{35-55}/CFA and observed for antigen-driven expansion by FACS analysis of blood at day 7 post immunization (dpi) (E) and clinical signs of EAE (n=6) (F). Each graph shows 1 representative of 3 independent experiments.

Loss of NIK Results in a Primary APC Defect

Our data and previous reports [103] support a T cell intrinsic role of NIK. However, we further aimed to identify the role of NIK in the accessory cell compartment. BMCs were generated by transferring a 4:1 mixture of $Rag1^{-/-}$ and $NIK^{aly/aly}$-2d2 BM into $Rag1^{-/-}$ mice. In those mice, $NIK^{aly/aly}$-2d2 T cell progenitors are developing within a NIK-sufficient accessory cell environment. Surprisingly, $Rag1^{-/-}+NIK^{aly/aly}$-2d2→$Rag1^{-/-}$ BMCs developed EAE comparable to $NIK^{aly/+}$-2d2→$Rag1^{-/-}$ BMCs (Fig. 19A). This finding demonstrates that NIK plays a vital role in hematopoietic accessory cells rather than in T cells to develop autoimmune inflammation. Complementing this result, polyclonally in vitro activated CD4$^+$ T cells from $Rag1^{-/-}+NIK^{aly/aly}$→wt BMCs were rescued in their ability to secrete IFNγ and Il-17 (Fig. 19B), which suggests that the production of effector cytokines by T cells requires intact NIK-signaling in the accessory cell compartment but not the T cell compartment.

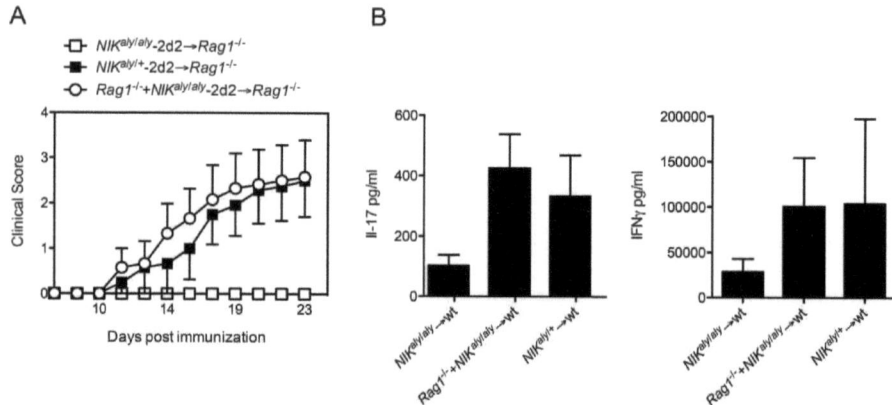

Figure 19: Loss of non-canonical NFκB-signaling in accessory cells determines the fate of T cells early in development. (A) Lethally irradiated $Rag1^{-/-}$ mice were reconstituted with BM of either $NIK^{aly/aly}$-2d2, $NIK^{aly/+}$-2d2 or a 4:1 mixture of $Rag1^{-/-}$ and $NIK^{aly/aly}$-2d2 mice. 6 weeks after reconstitution BMCs were immunized with MOG$_{35-55}$/CFA and observed for clinical signs of EAE (n=6). (B) Lethally irradiated wt mice were reconstituted with BM of either $NIK^{aly/aly}$, $NIK^{aly/+}$ or a 4:1 mixture of $Rag1^{-/-}$ and $NIK^{aly/aly}$ mice. 6 weeks after reconstitution CD4$^+$ T cells were isolated and stimulated with plate-bound α-CD3/α-CD28. Supernatants were analyzed by ELISA. Each graph shows 1 representative of 3 independent experiments.

DCs Maturation and Co-stimulation is Dependent on NIK-Signaling

The most prominent accessory cells involved in the induction of CMI and antigen-presentation are DCs. It has been reported that *in vitro* $NIK^{aly/aly}$ DCs have defects in the expression of CD80, CD86 and MHCII as well as in antigen-presentation and their ability to drive T cell expansion [123-125]. Also the ability of T cells to secrete effector cytokines is largely dependent on the capacity of APCs to provide T cell instructive cytokines. In particular, the cytokines Il-12, Il-23 and Il-6 have a major impact on the polarization of effector T cells. We investigated the ability of $NIK^{aly/aly}$ DCs to secrete these factors after stimulation with anti-CD40, thereby mimicking T cell-APC interactions. Interestingly, activated $NIK^{aly/aly}$ DCs secreted significantly lower levels of the pro-inflammatory cytokine subunit Il-12/Il-23p40 (Fig. 20A, B; Fig. 21A, B). We could further observe a strong reduction of the pro-inflammatory cytokine transcripts *Il-12p35*, *Il-23p19* (Fig. 21C) and protein levels of Il-6 (Fig. 20C). This data suggests that NIK-signaling in DCs upon DC-T cell interaction via CD40 is critical for the capacity of DCs to secrete T cell-instructive cytokines.

RESULT

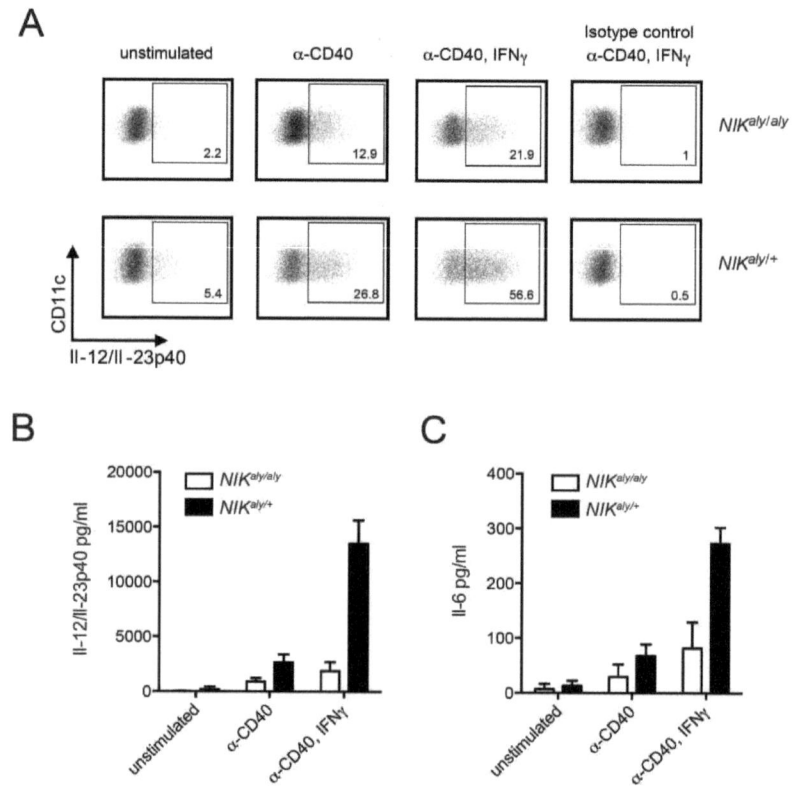

Figure 20: $NIK^{aly/aly}$ **DCs are restrained in secretion of pro-inflammatory cytokines. (A-C)** Splenic $NIK^{aly/aly}$ or control DCs were stimulated *in vitro* with α-CD40 and IFNγ for 24 hours. GolgiPlug was added for the last 4 hours of culture. Intracellular anti-Il-12/Il-23p40 FACS staining was performed together with cell surface staining for CD11c **(A)**. Supernatants were analyzed by ELISA for Il-12/Il-23p40 **(B)** and Il-6 **(C)**. Each graph shows 1 representative of 3 independent experiments.

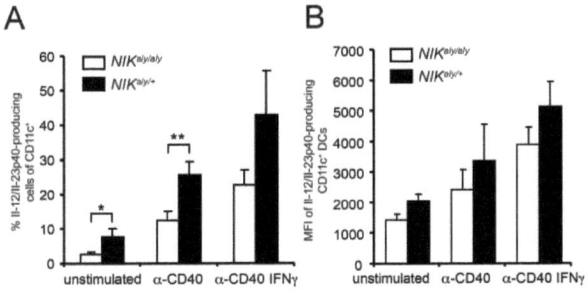

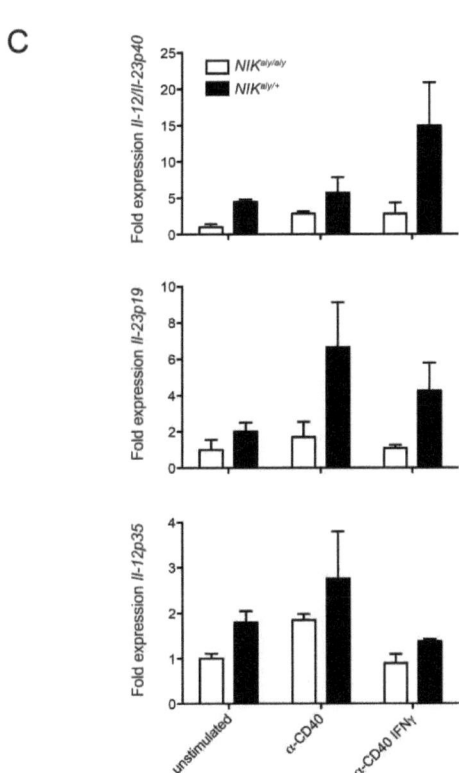

Figure 21: Splenic $NIK^{aly/aly}$ DCs were stimulated *in vitro* with α-CD40 and IFNγ for 24 hours. **(A-B)** GolgiPlug was added for the last 4 hours of culture. Intracellular Il-12/Il-23p40 antibody FACS staining was performed together with cell surface staining for CD11c. The percentage of Il-12/Il-23p40 producing DCs **(A)** and the *mean fluorescent intensities* (MFI) **(B)** are shown as means of ≥2 individual experiments. **(C)** Splenic $NIK^{aly/aly}$ DCs were isolated and stimulated *in vitro* with α-CD40 and IFNγ for 6 hours. Cells were harvested and whole RNA was isolated and reversely transcribed. qRT-PCR was performed to detect mRNA levels of Il-12/Il-23p40, Il-23p19 and Il-12p35. Shown are means of ≥2 independent experiments.

NIK$^{ALY/ALY}$ DCs are Restrained in T cell Priming and Fail to Induce EAE

In order to verify the reduced priming capacity of *NIK$^{aly/aly}$* DCs *in vivo*, we established a model based on diphtheria toxin (DTx)-mediated cell ablation, in which we generated a NIK-sufficient immune compartment where only DCs carry the mutated *NIK$^{aly/aly}$* protein. For this, a 1:1 mixture of *CD11cDTR* and *NIK$^{aly/aly}$* BM was transferred into irradiated wt recipients (Fig. 22A). Upon injection of DTx, DCs of *CD11cDTR* origin (*NIK$^{+/+}$*) are depleted while mutant *NIK$^{aly/aly}$* DCs are retained. The efficiency of DC-ablation was above 90% (Fig. 22B). We observed a significant delay in disease onset with decreased CNS infiltration of T cells, and particularly Il-17-producing T cells, in DTx treated *CD11cDTR+NIK$^{aly/aly}$*→wt BMCs compared to DTx treated control *CD11cDTR+NIK$^{aly/+}$*→wt BMCs (Fig. 22C, D, E). This finding strongly supports our previous *in vitro* and *in vivo* data and demonstrates that NIK-signaling in DCs is critical for their ability to instruct T cell polarization and effector function.

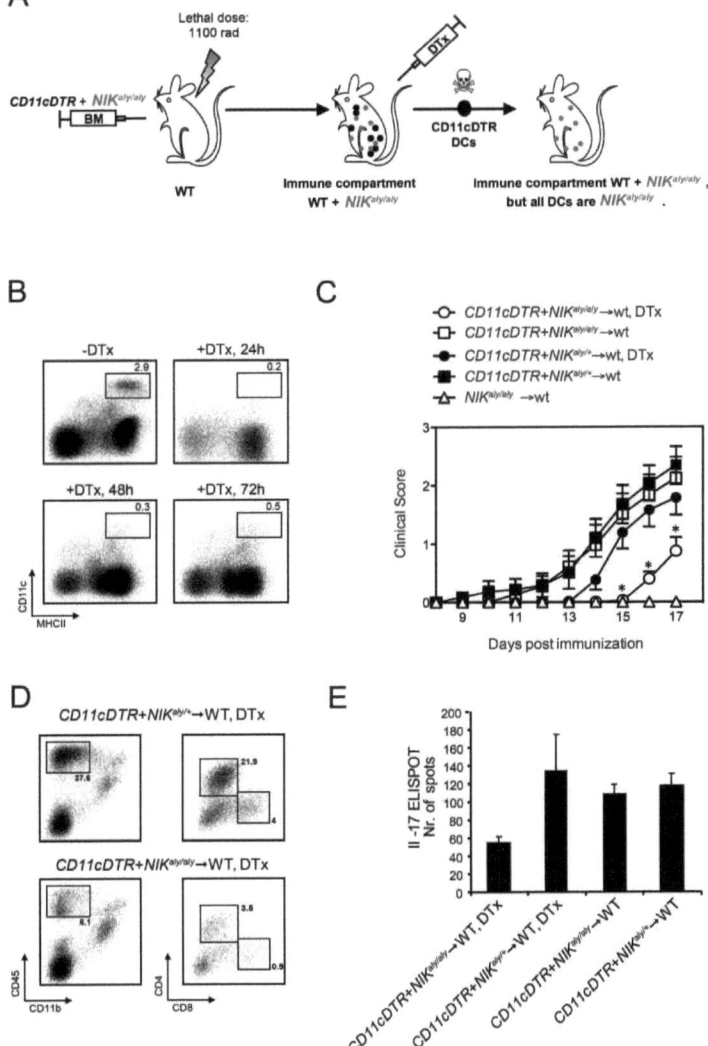

Figure 22: Absence of NIK-signaling in DCs significantly delays EAE. (A) Experimental scheme. Lethally irradiated WT mice were reconstituted with a 1:1 mixture of *CD11cDTR* and $NIK^{aly/aly}$ or $NIK^{aly/+}$ BM, respectively. Upon i.p. application of DTx, *CD11cDTR* ($NIK^{+/+}$) DCs are depleted, while $NIK^{aly/aly}$ or $NIK^{aly/+}$ DCs remain. All other immune cells are still present as $NIK^{+/+}$. **(B)** Efficiency of DC depletion in DTx-treated *CD11cDTR* mice was analyzed by flow cytometry. **(C)** *CD11cDTR+NIK$^{aly/aly}$*→wt and *CD11cDTR+NIK$^{aly/+}$*→wt BMC mice were immunized with MOG$_{35-55}$/CFA and treated with DTx every second day. Mice were observed for clinical signs of EAE. Three individual experiments were pooled (n≥15/group). **(D)** FACS analysis of CNS-infiltrating lymphocytes in MOG$_{35-55}$/CFA immunized DTx-treated *CD11cDTR+NIK$^{aly/aly}$*→wt and *CD11cDTR+NIK$^{aly/+}$*→wt BMCs at peak disease (17 dpi). Several brains and spinal cords of each experimental group were pooled. **(E)** Splenocytes were isolated from MOG$_{35-55}$/CFA immunized DTx-treated/untreated *CD11cDTR+NIK$^{aly/aly}$*→wt and *CD11cDTR+NIK$^{aly/+}$*→wt BMCs and rechallenged *in vitro* with 50µg/ml MOG$_{35-55}$ peptide, followed by Il-17 Elispot. Shown are triplicates of pooled splenocytes of one experiment (n≥5/group).

Restored NIK-Signaling in DCs But Not in T cells is Sufficient to Generate Pathogenic T cells

To ascertain a primary role of NIK-signaling in the DC compartment, we generated mice in which NIK expression is restricted exclusively to DCs. $R26Stop^{FL}NIK^{wt}$ mice express NIK^{wt} preceded by a loxP-flanked neo^R-Stop cassette and followed by a Frt-flanked *IRES-eGFP* within the ubiquitously expressed *ROSA26* locus [126]. Upon crossing to *CD11c-cre* mice, the neo^R-Stop is excised and NIK^{wt} will be expressed only in CD11c$^+$ cells, which are mainly DCs (these mice are hereafter called DC^{NIK}). Upon further breeding those animals onto the $NIK^{-/-}$ background, we generated mice that express NIK^{wt} in DCs, whereas all other cells and tissues completely lack the ability to express NIK.

In order to manipulate the expression of NIK only within the hematopoietic compartment and to provide a normal lymphoid organ structure, we again generated BMCs. We transferred BM of either DC^{NIK}-$NIK^{-/-}$ or control DC^{NIK}-$NIK^{+/-}$ and $NIK^{-/-}$ mice into lethally irradiated wt recipients. 6 weeks after reconstitution these BMCs were immunized with MOG$_{35-55}$/CFA. Strikingly, both DC^{NIK}-$NIK^{-/-}$→wt and DC^{NIK}-$NIK^{+/-}$→wt mice were fully susceptible to EAE, whereas control $NIK^{-/-}$→wt animals retained their EAE-resistance (Fig. 23A). Furthermore, the secretion of pro-inflammatory cytokines Il-12/Il-23p40 and Il-6 in NIK-sufficient DCs of DC^{NIK}-$NIK^{-/-}$→wt mice was largely restored (Fig. 23B). The exclusive presence of NIK in DCs rescued the secretion of Th effector cytokines Il-17, IFNγ and GM-CSF (Fig. 23C) after restimulation, regardless of the expression of NIK in T cells. These findings demonstrate that NIK-signaling only in DCs is sufficient to generate auto-aggressive T cells, even if these T cells themselves do not express functional NIK.

In contrast, when we expressed NIK exclusively in T cells by breeding the $R26Stop^{FL}NIK^{wt}$ mouse with *CD4-cre* and $NIK^{-/-}$ mice, T^{NIK}-$NIK^{-/-}$→wt BMCs did not develop any clinical signs of sickness (Figure 24). This experiment clearly demonstrated that the expression of NIK in T cells is elusive, whereas it is absolutely required in DCs.

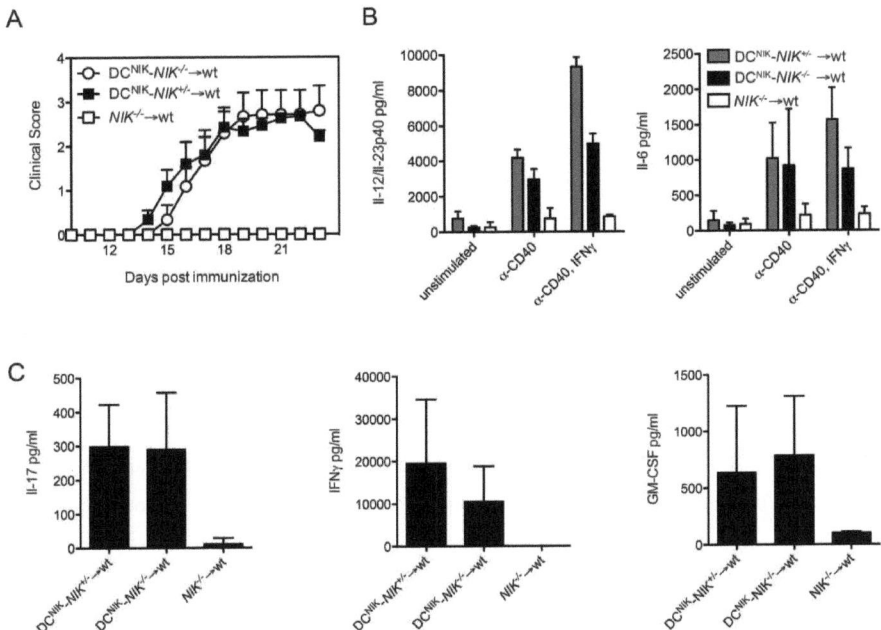

Figure 23: Expression of NIK in DCs is sufficient to restore EAE susceptibility in $NIK^{-/-}$ mice. (A) DC^{NIK}-$NIK^{-/-}$→wt and DC^{NIK}-$NIK^{+/-}$→wt BMCs were immunized with MOG_{35-55}/CFA and observed for clinical signs of EAE. Shown is one representative of 3 independent experiments (n=6). **(B)** Splenic DCs were isolated from DC^{NIK}-$NIK^{-/-}$→wt and DC^{NIK}-$NIK^{+/-}$→wt BMCs and stimulated *in vitro* with α-CD40 and IFNγ for 24 hours. Il-12/Il-23p40 and Il-6 secretion was measured by ELISA. **(C)** Splenocytes of MOG_{35-55}/CFA immunized DC^{NIK}-$NIK^{-/-}$→wt and DC^{NIK}-$NIK^{+/-}$→wt BMCs were isolated at peak disease and re-stimulated *in vitro* with MOG_{35-55} for 48 hours. Supernatants were analyzed by ELISA. Each graph shows 1 representative of 3 independent experiments.

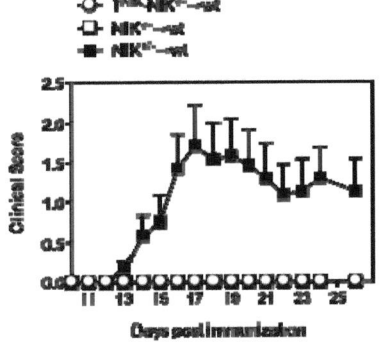

Figure 24: Expression of NIK in T cells is not sufficient to restore EAE susceptibility in $NIK^{-/-}$ mice. T^{NIK}-$NIK^{-/-}$→wt and $NIK^{+/-}$→wt BMCs were immunized with MOG_{35-55}/CFA and observed for clinical signs of EAE. Shown is one preliminary experiment (n=6).

LOSS OF NIK-SIGNALING CRITICALLY IMPAIRS THYMIC DCs FUNCTION

Our data thus far clearly show that NIK-signaling in DCs is critical for T effector function. However, if the restoration of NIK-signaling in DCs alone reinstates T effector function, the fact that adoptively transferred mature $NIK^{aly/aly}$ T cells into NIK-sufficient recipients fail to acquire effector function represents a contradiction. Hence, we hypothesize that NIK-signaling in thymic DCs is required to enable developing thymocytes to exit the thymus as fully functional T cells.

Therefore, the thymic DC compartment of $NIK^{aly/aly}$ and $NIK^{aly/+}$ mice was analyzed in detail. All three thymic DC subsets [127-129], namely migratory and resident conventional DCs (cDCs) ($CD11c^+CD172\alpha^+$ and $CD172\alpha^-$, respectively) and plasmacytoid DCs (pDCs) ($CD11c^{int}$ $CD45RA^+$) were found in comparable numbers in $NIK^{aly/aly}$ and $NIK^{aly/+}$ mice (Fig 25A). However, both cDCs and pDCs in $NIK^{aly/aly}$ mice expressed reduced levels of CD80, CD86 and MHCII (Fig. 25B), with the strongest reduction in the resident DC subset. The reduced levels of co-stimulatory molecules and MHCII indicate functional impairment of APC properties. To further investigate the phenotypic properties of $NIK^{aly/aly}$ thymic DCs we performed qRT-PCR analysis for various chemokine ligands and receptors involved in thymic DC function [128]. $NIK^{aly/aly}$ thymic DCs revealed a strong reduction in the expression of *CCL17*, *CCL19* and *CCL21* (Fig. 25D). The analysis of chemokine receptors further revealed an overall reduction in the levels of *CCR2*, *CCR5*, *CCR6* and *CCR7*, but increased expression of *CCR9* and *TLR9* (Fig. 26). To further assess their functional capacities, we co-cultured $NIK^{aly/aly}$ and $NIK^{aly/+}$ thymic DCs with 2d2 $CD4^+$ single positive (SP) thymocytes in the presence of MOG_{35-55}. $NIK^{aly/aly}$ thymic resident DCs elicited reduced proliferation in 2d2 $CD4^+$ SP thymocytes (Fig. 25C).

Taken together, we found that $NIK^{aly/aly}$ thymic DCs are phenotypically and functionally distinct from $NIK^{aly/+}$ thymic DCs, thereby having a possible negative impact on T cell development.

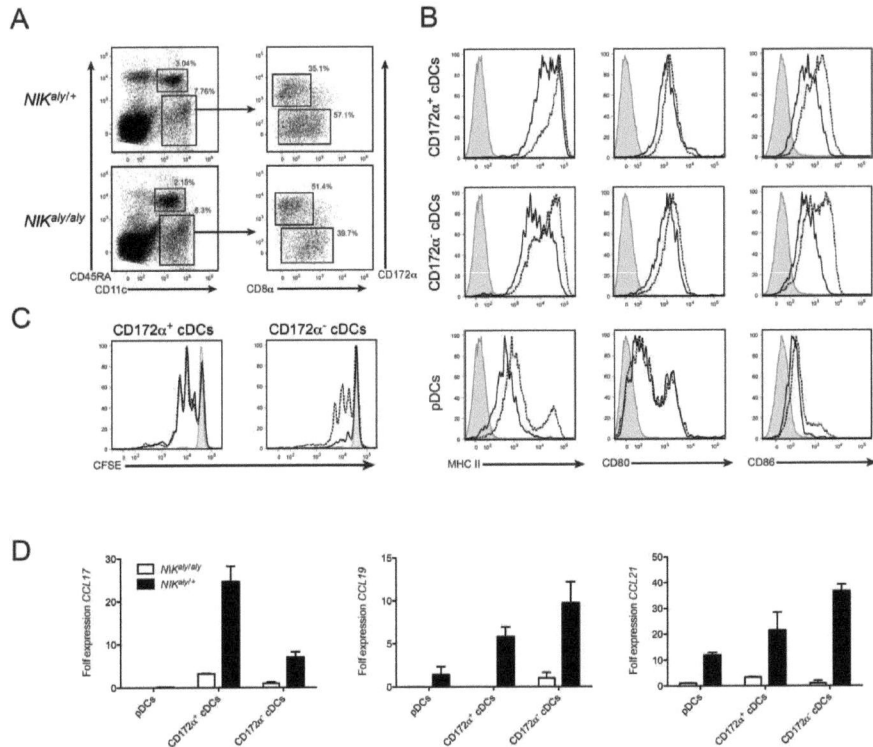

Figure 25: $NIK^{aly/aly}$ thymic DCs show reduced APC-capacity. (A) Flow cytometric analysis of thymic DC subsets from $NIK^{aly/aly}$ and control animals. pDCs are $CD11c^{int}CD45RA^+$, thymic resident DCs are $CD11c^+CD172a^-$ and thymic migratory DCs are $CD11c^+CD172a^+$. (B) Expression analysis of MHCII, CD80 and CD86 on thymic DC subsets. Dotted lines represent heterozygous control DCs, solid lines $NIK^{aly/aly}$ DCs and grey histograms unstained controls. (C) Proliferation of CFSE-labelled 2d2 $CD4^+$ SP thymocytes after three days of co-culture with thymic migratory (left) and resident (right) DCs in the presence of 10 µg/ml MOG_{35-55}. Dotted lines represent CFSE profile after stimulation with heterozygous control DCs, solid lines $NIK^{aly/aly}$ DCs and grey histograms show unstimulated cells. (D) RNA of FACS sorted thymic DC subsets from $NIK^{aly/aly}$ and $NIK^{aly/+}$ mice was transcribed into cDNA and analyzed by qRT-PCR for expression of different chemokines. Data is representative of 3 independent experiments.

RESULT

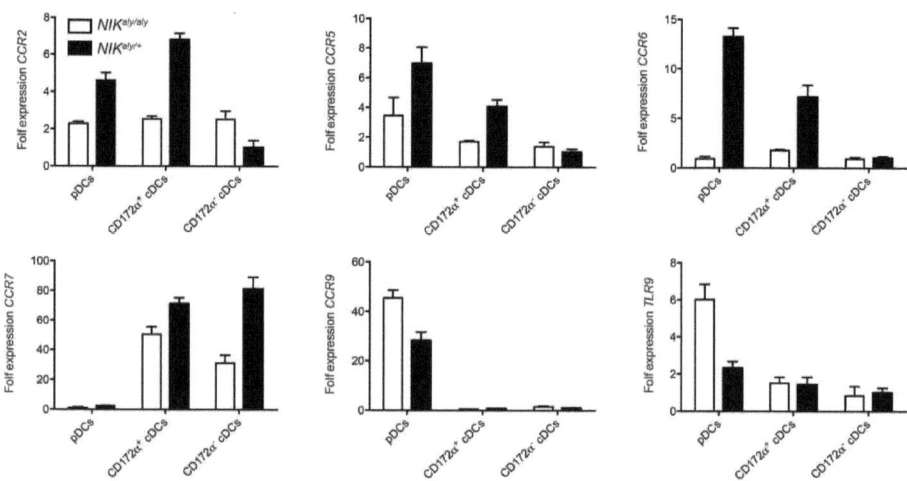

Figure 26: $NIK^{aly/aly}$ thymic DCs express a distinct pattern of receptors. RNA of FACS sorted thymic DC subsets from $NIK^{aly/aly}$ and $NIK^{aly/+}$ mice was transcribed into cDNA and analyzed by qRT-PCR for expression of different chemokine receptors and TLRs. Data is representative of 3 independent experiments.

RESTORATION OF NIK IN DCS RESCUES FOXP3, RORγT AND TBET EXPRESSION IN DEVELOPING THYMOCYTES

Thymic DCs have primarily been implicated in mediating negative selection [121]. However, there is increasing evidence that thymocyte development not only selects appropriate T cell receptors but also imprints effector function onto thymic emigrants. In particular, NIK has been suggested to be involved in the expansion of $CD25^+$ $CD4^+$ T cells [124, 130]. We found a 50% reduction in $FoxP3^+$ T_{regs} in both developing and mature T cells (Fig 27A, B). Given the reduced effector cytokine expression of $NIK^{aly/aly}$ T cells, we speculated whether the observed decrease of natural occurring nT_{regs} in $NIK^{aly/aly}$ thymi is the result of altered "licensing" of T cells during their development. Recently thymic T cell lineage commitment has been expanded to other lineages including T_H17 cells [131]. Therefore, we analyzed the gene expression of lineage-specific transcription factors in $CD4^+$ SP cells of $NIK^{aly/aly}$ thymi (Fig. 27C) and spleens (Fig. 27D) and found that also the expression of *RORγt* and *Tbet* was decreased, whereas *GATA3* was not affected. This suggests that the ability to become effector T cells including nT_{regs}, T_H1 and T_H17 cells is to some extent already determined during thymic development, and that thymic DCs are crucial for this process.

To verify that DC restricted expression of NIK is capable of restoring early lineage commitment, we again generated DC^{NIK}-$NIK^{-/-}$→wt and DC^{NIK}-$NIK^{+/-}$→wt BMCs and analysed the expression of transcription factors in CD4⁺ SP thymocytes. As shown in Fig. 27E and 27F, DC-restricted NIK expression fully restores *RORγt*, *Tbet* and also *Foxp3* expression.

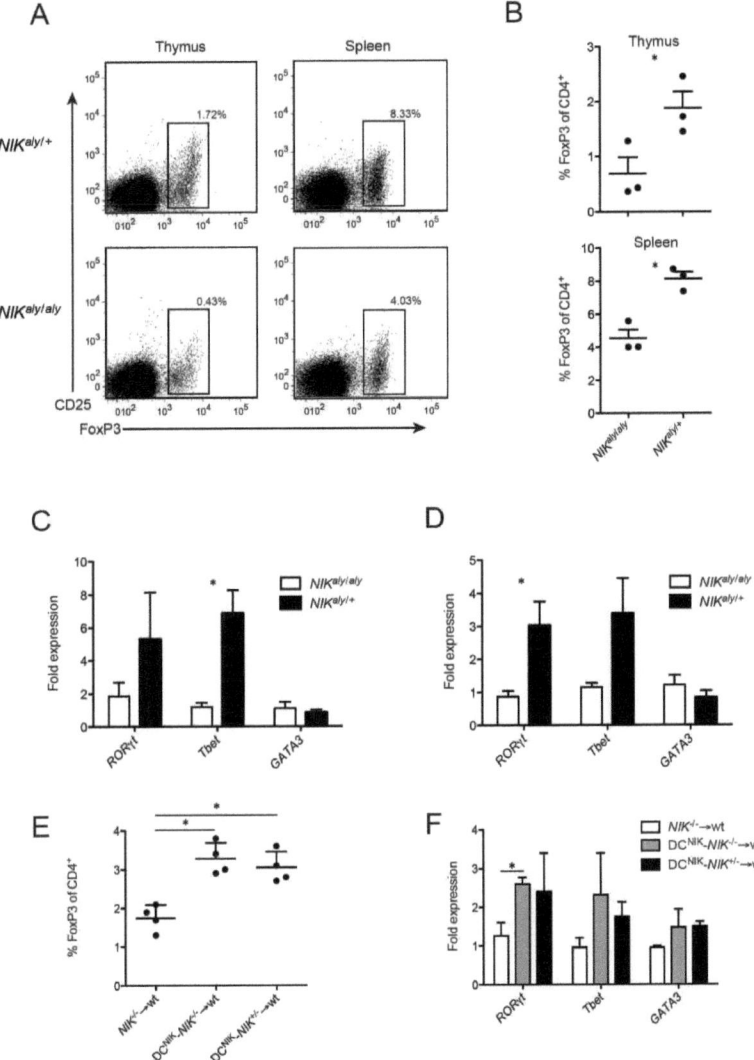

Figure 27: *Restoration of NIK in DCs rescues Foxp3, RORγt and Tbet expression in developing thymocytes*. (A-B) $NIK^{aly/aly}$ and $NIK^{aly/+}$ thymocytes and splenocytes were stained intracellularly for FoxP3 expression. Plots are gated on CD4⁺ CD8⁻ cells. (A). Shown are means of 3 mice each (B). (C-D) RNA of FACS sorted SP CD4⁺ thymic (C) and splenic (D) T cells of $NIK^{aly/aly}$ and $NIK^{aly/+}$ mice was transcribed into cDNA and analyzed by qRT-PCR for expression of *RORγt*, *Tbet* and *GATA3*. Data is representative of 3 independent experiments (n=3). (E-F) DC^{NIK}-$NIK^{-/-}$→wt, DC^{NIK}-$NIK^{+/-}$→wt and $NIK^{-/-}$→wt BMCs were generated. Percentage of FoxP3⁺ cells in the thymi (E) and transcript levels of *RORγt*, *Tbet* and *GATA3* in FACS sorted CD4⁺ SP thymocytes (F) were analyzed (n≥3).

Result

DISCUSSION

It is widely accepted that SLTs such as spleen and LN provide the basis for cooperative interactions between APCs, T cells and B cells [109]. Whereas the spleen's major role lies in the protection against blood-borne infectious agents [132-134], LNs drain antigens from local sites of infections. SLTs are highly organized structures composed of well defined B cell and T cell zones, which bring APCs, T- and B cells in close proximity to efficiently enable scanning for the cognate antigen [135]. The notion that SLTs are required for the initiation of adaptive immune responses is not only supported by the swelling of LNs after a local infection, but also by the immunodeficiency of alymphoplastic mice which are genetically deficient in lymph nodes. Such mice were frequently used to underline the importance of SLTs in forming adaptive immune responses. However, there are a number of reports, which show that immunization-induced expansion of T-lymphocytes can occur outside of SLTs, while B cell maturation including the production of Abs is highly dependent on the topography in SLTs. I will aim to summarize studies, in which the resourcefulness of immunity prevails over structural constraints of such SLTs. Much of the observed phenomena discussed here can be explained by investigating the reliance of immunity on SLTs during the evolution from primitive jawed fish to complex mammals and the implications for our current understanding of mammalian immunology.

Further, upon revealing that the lack of SLTs is not the main cause of the immunodeficiency of $NIK^{aly/aly}$ mice, I will discuss the requirement of NIK in adaptive immunity and more precisely in CMI. This discussion will lead us to novel mechanisms, which facilitate thymic DCs to license the T cell repertoire early during development.

Discussion

THE ROLE OF SLTs FOR THE DEVELOPMENT OF CELL-MEDIATED IMMUNITY VERSUS HUMORAL IMMUNE RESPONSES

CAN PRIMARY IMMUNE RESPONSES ONLY BE INITIATED IN SPECIALIZED LYMPHOID STRUCTURES?

The formation of an immunological synapse has been studied in great detail at the molecular level. For instance immunological synapses form between APCs and T cells *in vitro*. Both T- and B cells can be primed *in vitro* using receptor-crosslinking with monoclonal antibodies (mAbs) or artificial APCs [136, 137]. Hence, such artificial microenvironments have not only enabled us to determine the molecular underpinnings of T- and B cell priming, but also demonstrate that the minimal requirements for priming are determined by molecular interactions, rather than the structural environment.

Nevertheless, for decades it has been postulated that the induction of adaptive immunity occurs in organized lymphoid organs [138]. Studies in mice lacking organized lymphoid tissues (alymphoplastic) have emphasized the pivotal role of SLTs in the initation of an immune response. NIK$^{aly/aly}$ mice lack LNs and PP and show disrupted architecture of the spleen and thymus [94, 99]. The NIK$^{aly/aly}$ phenotype is caused by a point mutation within the NIK, an important mediator of the alternative NFκB-pathway, which plays a major role in the development of the embryonic *anlage* of SLTs [99, 139]. NIK transduces signals from the TNFR family that leads to the processing and cleavage of p100 to p52 [123, 139]. Signaling through the TNFR family is involved in the development of SLTs via LTβ-signaling, B cell maturation (via BAFF), and adaptive immune responses (via CD40) [64].

Before the role of NIK in adaptive immunity was suggested, NIK$^{aly/aly}$ mice were frequently used as an experimental model to demonstrate the importance of SLTs for the induction of primary immune responses. The first such discovery was the inability of these mice to reject allogeneic skin grafts, and was attributed to the lack of SLTs even before the origin of the NIK$^{aly/aly}$ mutation was revealed [94]. The deficit of NIK$^{aly/aly}$ mice to fight viral infections such as lymphocytic choriomeningitis virus (LCMV) was explained by the insufficiency of T cell responses due to the lack of LNs [109]. Similarly, the fact that NIK$^{aly/aly}$ mice cannot reject tumors was explained by the fact that tumor cells or tumor Ags had no LNs to migrate to and were thus unable to induce a CTL response [100]. Another group showed that splenectomized NIK$^{aly/aly}$ mice were not able to reject cardiac allografts [101], again, owing to the lack of SLTs. LTα$^{-/-}$ mice, which also lack LNs and have a similar phenotype to NIK$^{aly/aly}$ mice, do not elicit a contact-hypersensitivity response to epicutaneous hapten-immunization, further indicating that LNs are required for induction of CMI [108]. Finally, much like NIK$^{aly/aly}$ mice, LTα$^{-/-}$ mice show greatly reduced LCMV-specific CTL responses as well as abnormal CTL effector functions to herpes simplex virus (HSV) [140, 141]. Collectively, these results seem to suggest that primary immune responses to a whole variety of triggers, whether alloantigens, viruses, or tumors depends on the presence of normal SLT architecture.

HAS THE IMMUNODEFICIENCY IN NIK$^{ALY/ALY}$ MICE BEEN PROPERLY INTERPRETED?

Apart from the absence of LNs, NIK$^{aly/aly}$ mice host a variety of other serious immune defects such as reduced immunoglobulin (Ig) production and isotype class switch, defective T cell function and homing responses [94, 105, 106, 142]. Until now, it has been somehow neglected that these defects could potentially be connected to the role of NIK in cell activation rather than the developmental defects. Here, I summarize different mouse models used to study the importance of SLTs to primary immune responses, and present evidence that potent T cell responses can in fact occur independent of SLTs.

Several groups showed that NIK$^{aly/aly}$, and the almost equivalent NIK$^{-/-}$ mice are not only immuno-compromised due to the lack of SLTs but also due to the loss of function mutation within NIK in immune cells. Apart from the vital role of NIK in LTβR signaling and LN formation, NIK has been implicated in a variety of cellular processes. It was shown that NIK-deficient T cells produce less Il-2 and GM-CSF, and have proliferation deficits upon TCR-engagement [70, 102, 105]. Furthermore, recent reports demonstrated that NIK$^{aly/aly}$ T cells cannot differentiate into Th17 cells and fail to become auto-pathogenic when presented with cognate self-antigen [103, 104]. By reconstituting wt mice with an NIK$^{aly/aly}$ hematopoetic system, we demonstrated that the failure to develop pathogenic T cells is not connected to the lack of SLTs but lies exclusively in the NIK-mutation in immune cells. In addition, we and others demonstrated the inability of NIK-deficient T cells to induce EAE and to develop into cytokine-secreting effector cells [103, 104]. Furthermore, several reports provided evidence that NIK-signaling is not only crucial for T cells but also for professional APCs. For example NIK$^{aly/aly}$ DCs have been shown to express low levels of MHCII and costimulatory molecules such as CD80, CD86 and show deficits in antigen-presentation to and activation of CD4$^+$ T cells, including also the expansion of regulatory T cells [124]. NIK$^{aly/aly}$ DCs further display a reduction of CD40-mediated Il-12 production [143] and fail to cross-prime CTLs to exogenous antigen, seemingly due to multiple defects in antigen-processing [125].

However, after reconstitution of NIK$^{aly/aly}$ mice with a wt hematopoetic system and further splenectomy, mice have been shown to mount pathogenic T cell responses proving clearly that CMI can be induced in the absence of SLTs [104]. The observation that the distinct structure of SLTs does not seem to be critical for the induction of CMI was underscored in studies using models other than NIK$^{aly/aly}$ mice, but which also lack or have otherwise disrupted SLTs (summarized in Table 2). More evidence that SLTs are not vital for the development of CMI stems from the findings that mice lacking LTβR and hence LNs are capable of mounting strong adaptive immune responses to neo- as well as auto-antigen [104, 144]. In addition, LTα$^{-/-}$ mice, which also lack LNs, are able to mount potent CMI [145] and they have been shown to generate even stronger responses than wt mice in an airway-inflammation model [146]. Upon intranasal infection with influenza, LTα$^{-/-}$ mice were competent to generate virus-specific CD8$^+$ T cells that produced IFNγ and exhibited killer-activity [115]. Similarly to influenza infection, upon intranasal application, LTα$^{-/-}$ mice cleared a productive murine gammaherpesvirus 68 (MHV-68) infection and controlled latent infection with CTL-activity comparable to wt mice, although with delayed kinetics [147]. Furthermore, LTα$^{-/-}$ mice were able to reject allogeneic small bowel transplants and skin allografts [148, 149]. Another group has described effective T cell priming and rejection (albeit in an impaired fashion) of allografts in splenectomized LTα$^{-/-}$ and LTβR$^{-/-}$ mice, indicating that SLTs contribute but are not absolutely required for immune responses in transplantation settings [144]. Also upon gastrointestinal infection with the gut-residing protist parasite *Eimeria vermiformis*, LTα$^{-/-}$ mice recruited antigen-specific Th1 cells to the gut, although this was delayed compared to wt mice with intact mesenteric LNs (mLN) and PP [150]. We will discuss below how some of the above-described observations (i.e. responses upon lung and gut infections) can potentially be explained by the fact that local inflammatory stimuli cause the development of TLTs.

More evidence, that CMI can be initiated in the absence of SLTs, comes from so-called *plt* mice. *Plt* mice lack the chemokines CCL21 and CCL19 expression in SLTs and are therefore characterized by a defect in the homing of naïve T cells and DCs to LNs, PPs, and splenic white pulp [151-154]. It was demonstrated that *plt* mice are capable of mounting robust CD4$^+$ T cell responses upon s.c. immunization with OVA peptide and contact-sensitization [155, 156]. Those T cell responses in *plt* mice were even enhanced compared to wt mice and did not decline for the normal contraction-phase, suggesting that the structure of SLTs, at least for subcutaneously delivered antigens, is more important for the elimination of effector T cells than the actual priming and expansion-phase. Furthermore, *plt* mice could demonstrate delayed but pathogenic CD4$^+$ T cell-mediated autoimmune disease when presented with self-antigen plus adjuvant [104]. Similar findings showed that in an infectious model using LCMV, CTLs in *plt* mice could also be primed efficiently outside of the discrete T cell zones of SLTs [120].

Table 2: Efficient T cell responses in mouse models, which lack or have disturbed SLTs.

Mouse model	T cell response	Reference
LTα$^{-/-}$ mice	Normal cytotoxicity of CD4$^+$ and CD8$^+$ T cells	[145]
	Strong airway inflammatory response	[146]
	Sufficient CD8$^+$ T cell response upon i.n. influenza infection	[115]
	Efficient CD8$^+$-mediated clearance of MHV-68	[147]
	Rejection of intestinal transplants	[148]
	Rejection of skin-allografts	[149]
	CD4$^+$-mediated rejection of heart- and skin-allografts in splenectomized LTα$^{-/-}$ mice	[144]
	T cell recruitment to the gut upon *E. vermiformis* infection	[150]
	Robust primary and secondary T cell response to influenza	[114, 118]
LTβR$^{-/-}$ mice	Development of EAE	[104]
	CD4$^+$-mediated rejection of heart- and skin-allografts in splenectomized LTβR$^{-/-}$ mice	[144]
Plt mice	Robust CD4$^+$ T cell responses upon contact-sensitization	[155]
	Robust CD4$^+$ T cell responses upon OVA/CFA immunization	[156]
	Development of EAE	[104]
	Induction of LCMV-induced antiviral CD8$^+$ T cell responses	[120]
NIK$^{aly/aly}$ mice	Development of EAE in splenecomized NIK$^{aly/+}$ → NIK$^{aly/aly}$ bone marrow chimeras, efficient tumor rejection upon vaccination	[104]
	Proliferation and differentiation of effector and memory T cells in the bone marrow of splenectomized NIK$^{aly/aly}$ mice	[117]
Others	T cell priming in the bone marrow of Mel-14 treated splenectomized mice	[157]

Definitions: NIK$^{aly/aly}$, alymphoplastic; CFA, complete Freund's adjuvant; EAE, experimental autoimmune encephalitis; i.n., intra-nasal; LCMV, lymphocytic choriomeningitis virus; LTα, lymphotoxin alpha; MHV-68, murine herpes virus; OVA, ovalbumin peptide

IN CONTRAST TO T CELL ACTIVATION, B CELL ACTIVATION AND CLASS SWITCHING DEPENDS STRICTLY ON INTACT LYMPHOID STRUCTURES

As discussed above, there is accumulating data suggesting that SLTs are not an absolute requirement for the induction of CMI. However, in striking contrast different rules seem to apply for the induction of B cell mediated humoral immunity. Neither LTα$^{-/-}$, LTβR$^{-/-}$, nor NIK$^{aly/aly}$ mice generate IgA Abs to a variety of triggers [94, 96, 97]. Similarly, upon s.c. infection with inactivated herpes simplex virus-1 (HSV-1), LTα$^{-/-}$ mice do not produce HSV-1-specific IgG Abs [96]. LTβR$^{-/-}$ mice also show strongly reduced IgG responses and affinity maturation to immunization with 4-hydroxy-3-nitrophenol-acetyl-chicken gamma globulin (NP-CG) in alum [97]. LTα$^{-/-}$ mice immunized s.c. with sheep red blood cells (SRBCs) mount a strong IgM response but fail to switch to IgG [158]. Adoptive transfer studies of wt splenocytes into LTα$^{-/-}$ mice or the reciprocal, showed that the failure of Ab class switching was not B cell intrinsic but rather due to the lack of GCs and FDC clusters in LTα$^{-/-}$ spleens [158]. Furthermore, the LTα$^{-/-}$ spleen is even insufficient to provide

DISCUSSION

the structural needs for functional memory IgG responses after adoptive transfer of sensitized memory cells from wt mice followed by antigenic challenge, showing that splenic structure is also required to permit an isotype-switched memory Ig response [116]. Similar results have been observed by s.c. immunizing LTα$^{-/-}$ mice that have been reconstituted with a wt immune compartment [104]. Although IgM levels were comparable to controls, mice were not able to perform Ab class switch to IgG. Conversely, upon intranasal infection with MHV-68, LTα$^{-/-}$ mice mounted high titers of IgM and different IgG classes [147], suggesting that the manner of antigen administration does matter. This observation was confirmed by two studies showing that upon intranasal infection of splenectomized LTα$^{-/-}$ mice with influenza virus elicited virus-specific IgM and IgG Abs [115, 118]. Again, this phenomenon can be explained by the emergence of TLTs within the lung and will be discussed below.

The inability of NIK$^{aly/aly}$ mice to generate productive Ab-responses has been described thoroughly. Just like LTα$^{-/-}$ and LTβR$^{-/-}$ mice, NIK$^{aly/aly}$ mice lack GC-formations and are defective in both Ig isotype-switch and hypermutation [95]. Upon infection with vesicular stomatitus virus (VSV) or vaccinia virus (Vacc), NIK$^{aly/aly}$ mice displayed delayed IgM responses and failed completely to generate virus-specific IgG [109]. NIK$^{aly/aly}$ B cells are largely unresponsive to CD40-induced proliferation, Ig production and NF-κB activation, suggesting an intrinsic defect [123]. The NIK$^{aly/aly}$ B cell intrinsic defect was confirmed by the NIK$^{aly/aly}$ bone marrow (BM) transfer into wt mice, which also resulted in poor B cell follicles and lack of isotype-switching even in the presence of SLTs and FDC clusters in the spleen [102]. However, we have recently shed light on the issue of whether SLTs are an absolute requirement for the induction of humoral immunity. By the transfer of wt BM into NIK$^{aly/aly}$ mice and splenectomy we generated a model that lacks all SLTs but contains a functional wt immune system [104]. Upon s.c. antigen-administration these mice failed to generate a productive Ab response or class switch. This finding was supported by a study showing that Vacc-specifc IgG responses could not be restored with an adoptive transfer of wt splenocytes into NIK$^{aly/aly}$ mice [109].

Taken together, the above data suggests that upon respiratory viral infections, specialized lymphoid structures in the lung (TLTs) can be assembled upon inflammation independent of LTα-signaling and provide the structural needs to B cells in order to allow class switch and memory formation. However, if conventional SLTs are absent and during different infection routes, antigen will not lead to inflammation in the lung. The absence of specialized lymphoid structures will then lead to insufficient B cell priming and maturation. In spite of the fact that some evidence exists that splenic autoreactive B cells can expand and undergo somatic hypermutation at the T zone-red pulp border and not in GCs [159], most observations indicate clearly that B cells are dependent on lymphoid structures composed of FDC clusters and GCs.

In spite of the widely held belief that adaptive immunity in general requires the structural support of SLTs, there is ample evidence to indicate that only B cells, but not T cell are dependent on SLT. We will next discuss the basis for these phenomena by exploring the development of dedicated SLTs during the evolution of the immune system in vertebrates.

THE EVOLUTION OF LYMPHOID TISSUES AND THE IMPACT ON ADAPTIVE IMMUNITY

Vertebrates consist of two subphyla: that of jawless (Agnatha) and that of jawed (Gnathostomata) vertebrates. Both subphyla have developed an adaptive immune system based on somatic random generation of a clonally expressed repertoire of lymphocyte receptors. These adaptive antigen-receptors are expressed by cells of mesodermal origin, i.e. the lymphocytes, whose development shares a network of transcription factors [160]. The random combination of scattered genetic elements (Leucine-rich repeats in Agnatha and Immunoglobulin Superfamily (IgSF) in Gnathostomata) is the source of a large diversity of receptors and Abs [161]. Recent studies showed that the variable lymphocyte receptors (VLRs) of Agnatha (i.e. in lampreys) are even more diverse and efficient in antigen recognition than originally thought [162-165]. The Gnathostomata Ab repertoire generated by RAG-mediated rearrangement of IgSF gene segments can be further modified by somatic mutation, class switch and gene-conversion, which involve an activation-induced deaminase (AID) Apobec family member of cytidine deaminase, which is already present in Agnatha [162, 163]. AID-mediated class switch can be found from tetrapods (amphibians) onwards [161]. Whatever the analogies and convergences between Agnatha and Gnathostomata, there are still major differences between the two subphyla and this review addresses the features more specific of the Gnathostomata (IgSF receptors, affinity maturation, somatic mutation, class switch, memory formation). Once precise data on the lymphoid tissue of the Agnatha will be available it will be interesting to see to which level the convergence can be extended.

B- and T-lymphocytes emerged simultaneously during phylogeny in jawed vertebrates. At the molecular level of the receptors, no major element seems to be missing in non-mammalian vertebrates for developing a mammalian-type of immune response i.e. with high-specificity and memory. Yet there are differences. Are they linked to the organization of the lymphoid organs? As discussed above, in contrast to T cells mammalian B cells have become accustomed to organized lymphoid organs, whose physical organization shows a rare example of a progressive complexity within the phylum [166]. The basic evolution of lymphoid organs in relation to various aspects of the immune responses are presented in Figure 28. We will here discuss whether the evolution of such structures coincides with the potent affinity maturation seen only in mammals.

The complexification of SLTs might be responsible for the increased efficiency of the immune responses and the selection events associated with it seen as one moves from ectotherms to endotherms. Increased efficiency does not mean that the response of ectotherms lack completely efficiency. We are talking about a hierarchy in efficiency and selection. Clearly without selection at all there would be no adaptive immune response.

Within the context of this review we shall not cover the conserved primary lymphoid organs: the thymus and the BM or their equivalents, which remain unidentified in Agnatha but are well characterized in all Gnathostomata. We shall focus on SLTs of Gnathostomata, where T- and B cell responses are initiated.

DISCUSSION

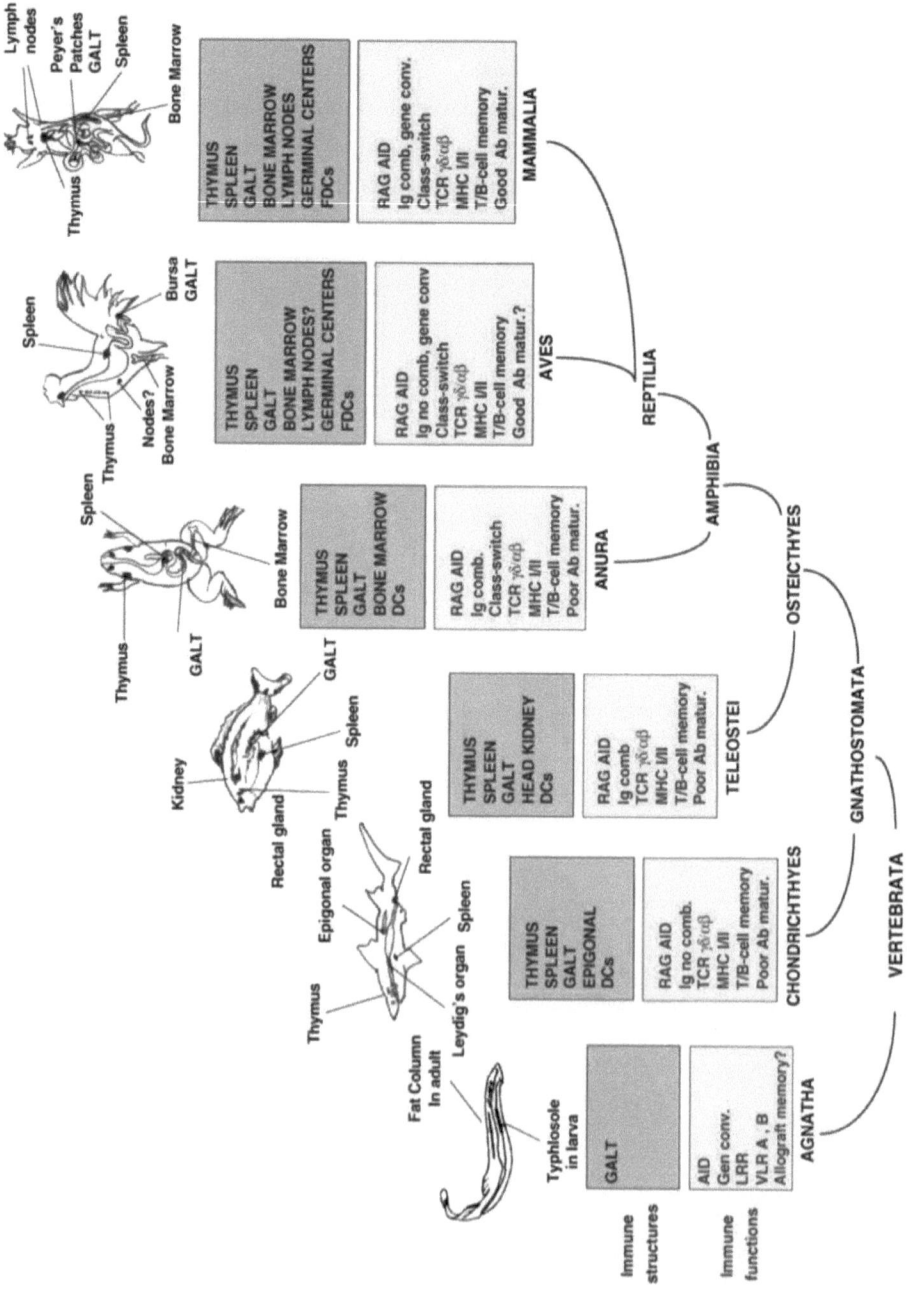

Figure 28: Lymphoid organs and immune responses across vertebrates
The figure describes the complexification of lymphoid organ architecture and number across Gnathosotmates in parallel with immune function parameters.
Abbreviations: AID, Activation induced cytidine deaminase; Gen conv, gene conversion; LRR, Leucine rich repeat; VLR, Variable lamprey receptor; RAG, recombination activating gene; Ig comb, possibility of combinatorial association of V(D)J elements; TCR, T cell receptor; Ab matur, antibody affinity maturation; GALT, gut-associated lymphoid tissue; DC, dendritic cell; FDC, follicular dendritic cell; MHC, Major histocompatibility complex
Adapted from the following references [161], [166], [167],[168], [169], [170], and [171].

LYMPHOID STRUCTURES IN VERTEBRATE EVOLUTION

THE SPLEEN IN VERTEBRATE EVOLUTION

With its lymphoid white and hematopoietic red pulps the spleen is present in all gnathostomes. It filters the blood, removes damaged cells and plays an important role in immunity towards blood-borne antigens. Cartilaginous fish spleen contains well-defined, highly vascularized white pulp areas, with a central T cell zone containing a MHC class II$^+$ DC network and Ig secretory cells, surrounded by smaller zones of B cells [172].

Bony fish spleen, despite a poor development of the white pulp, contain the basic elements of the mammalian spleen but few cells resembling FDCs. Special terminal capillaries, the ellipsoids, have a thin endothelial layer surrounded by fibrous reticulum and with a sheath of macrophages that are involved in antigen-trapping. Lymphocytes are seen in their vicinity during immune responses. Consequently, ellipsoids have sometimes been interpreted as primitive GCs.

In anuran amphibians (i.e. frogs) the white pulp is organized in follicles with reticular elements, macrophages, DCs, T- and B cells. T cells are found at the periphery of the follicles and B cell zones expressing AID [173] are central and fitted around the central artery. No GCs have been observed in any amphibians.

In reptiles, the white pulp shows a higher degree of compartmentalization. Its periarteriolar lymphoid sheath (PALS) contain small lymphoblasts, phagocytes, whereas the periellipsoid lymphoid sheath (PELS) contain medium-sized lymphocytes and cells resembling FDCs but canonical GCs are absent [166].

Bird spleen resemble that of reptiles more than that of mammals [174]. It contains 3 distinct areas: PALS, PELS, and follicles with GCs. Birds are the earliest vertebrates to show this specialized structure.

In addition to the above-mentioned conserved features, the mammalian white pulp nodules are surrounded by a marginal zone rich in macrophages (histoenzymatically similar to the ellipsoid-associated macrophages of fish), where antigen gets trapped. Follicles of the mammalian spleen contain GCs [166].

DISCUSSION

THE GUT-ASSOCIATED LYMPHOID TISSUES IN VERTEBRATE EVOLUTION

Gut-associated lymphoid tissue (GALT) is present in all vertebrates and in fact the presence of lymphocytes in the gut, an area particularly exposed to pathogens, has suggested the conservation of a common mucosal immune system for all vertebrates. "GALT" is essentially a geographical denomination and the cells and structures found there may differ in their significance. Lymphoid aggregates seen in mesentery and the lamina propria are frequent in cold-blooded vertebrates but do not represent well-organized lymphoid organs [166]. The synthesis of IgX, an analog of IgA, is secreted by plasma cells in the gut-epithelium of *Xenopus* whereas many T cells are found in the lamina propria. They most likely correspond to SLTs engaged in responses [175]. In fish a highly diversified alpha-beta TCR repertoire is maintained in intraepithelial lymphocytes in the absence of PPs and mLNs [176].

PPs are seen only in birds and mammals. They appear as oval or round lymphoid follicles (similar to LNs) located in the lamina propria layer of the mucosa and extending into the submucosa of the ileum. In adults, B-lymphocytes predominate in the follicles' GCs, whereas T cells are located between follicles. In ruminants the large ileal PP functions in fact as a bursa-like primary lymphoid organ [177].

LYMPH NODES IN VERTEBRATE EVOLUTION

There are no LNs in fish, and none of the lymph glands, cavity bodies, jugular bodies of amphibians have a LN architecture. They are blood-filtering organs without any connection to lymph. However, in reptiles (lizard) some aggregates form nodules associated with the lymphatic network resembling LNs of primitive mammals such as the echidna [166].

Duck "lymph nodes" have been described but there are very few of them and their structure is different from that of mammals even though they show GCs and post capillary venules for recirculation [178]. By definition, LNs with a canonical architecture are not seen before mammals, where they can be very abundant (e.g. 500 to 600 in humans). They are particularly efficient during responses against antigens present in the specific tissue thanks to their lymphatic vessel system. They consist of a cortical region with primary follicles rich in B cells and FDC. T cells and DCs are found in the paracortex. Macrophages and plasma cells reside in the medullary region of the node. The compartmentalization of the cells within the node is under the control of a complex network of chemokines [179].

DISCUSSION

In summary, while all jawed vertebrates have true SLTs, ectotherms lack LNs and organized GALTs. GCs with many FDCs are not formed after immunization until the phylogenetic position of birds that are also the first vertebrates possessing LN-like architecture, although not at all to the same extent as mammals, which stand out in terms of the organization of their SLTs and their ability to generate high affinity Igs.

THE IMMUNE RESPONSES OF JAWED VERTEBRATES WITH A SPECIAL EMPHASIS ON COLD-BLOODED VERTEBRATES

For a long time it has been claimed that virtually no affinity maturation took place in fish, amphibians, or reptiles. The existence of immunological memory has also been questioned, especially in fish. Following the discovery of hypermutation in cartilaginous fish, immune responses have been reinvestigated *in vivo*. Following an injection of a T-dependent antigen, the affinity of the monomeric IgM and an unusual heavy chain homodimeric isotype IgNAR in sharks (*Ginglymostoma cirratum*) increases about 10-fold over 7-9 weeks [180]'[181]. A quicker secondary response can also be demonstrated but without an increase in titers over those in the primary response. The production of the monomeric IgM, a subclass of shark heavy chain [182], is probably not due to a switch but due to the involvement of distinct cell lineages committed to one isotype or the other [183] and not yet fully understood. In the more recently evolved bony fish (teleosts) such as the trout, immunological memory is demonstrable either by the success of some vaccines or by immunization with sheep red blood cells as antigen. The anamnestic character (i.e. faster response and higher Ab titers) of the secondary response against this T-dependent antigen was demonstrated at 24° C and 20° C but lost at 18° C [184]. T cell functions are notoriously impaired at low temperature [185]. Minor increases in affinity maturation in the trout are attributed to selections of either minor pre-existing B cell populations or of somatic mutants. So far, no switch has been reported to any of the newly discovered non-IgM isotypes. Isolation of TCR genes and the existence of polymorphic MHC I and II molecules in teleosts suggest that antigen presentation is operative. Indeed, following an acute infection with viral hemorrhagic septicemia virus, complementary-determining region (CDR)-specific T cell clonal expansion and TCR CDR3 profiles were skewed for several $V\beta/J\beta$ combinations [186].

Among anuran amphibians (i.e. frogs), *Xenopus* responds specifically to many T-dependent antigens, with restricted Ab heterogeneity and a modest (4-10 fold) affinity maturation following immunization with the nominal antigen DNP-KLH. The memory response peaks at 2 weeks instead of 3-4 and is 10 to 40-fold higher, but there is no further change in affinity over the initial rise.

Somatic mutations occurring at hypermutation rates in the VH of B cells from immunized animals, were not strongly localized in CDRs [187]. Thus paradoxically, there is in principle no shortage of variants, and the reasons for the poor Ab affinity maturation might be a less than optimal selection in the absence of GCs. Yet, although affinity maturation does not appear to occur, isotype-switch to IgY is observed. This depends on T cells and does not occur at low temperatures. Other essential T cell functions in immune responses were demonstrated with thymus-dependent allograft rejection studies and with *in vitro* assays for T–B collaboration and MHC restriction. Histologically, Ab-forming cells can be observed scattered in the white pulp of the spleen or in proliferation foci in the liver both of *Alytes* larvae and in *Xenopus* larva and adults [188, 189], where they might correspond to the genesis of TLTs [190]. Urodele amphibians such as the newt elicit low IgM Ab heterogeneity and they respond poorly to thymus-dependent antigens even though their expressed TCR diversity looks normal. This might also be related to the fact that the urodele spleen is not clearly divided into white and red pulp. Much less data are available in reptiles although the lack of an increase in affinity and low Ab heterogeneity has been reported [191, 192]. There is currently no data on T cell effector functions of reptiles during an immune response.

In birds, chicken Ig sequence data and light chain patterns on 2-D gel electrophoresis showed less Ab heterogeneity than the mouse. A poor increase in affinity maturation of chicken Ab responses has been reported. However ducks display Ab responses, affinity maturation (ca. 400 fold) and memory to immunization with TNP-KLH or TNP-HIgG (both T-dependent antigens) and adjuvants [193]. These results could be explained with the existence of non-conventional LNs in the duck whereas these organs have not been found in chickens.

In conclusion, although all Gnathostomata have the potential for generating diverse Abs after immunization, only homeothermic vertebrates such as mammals make the most of this potential through the process of efficient affinity maturation. Besides the phylogenetically universal T-dependent allograft rejection; memory, specificity, and some comparatively modest affinity maturation, have now been demonstrated in fish and amphibians, yet often only with difficulty. Somehow the immune responses of ectotherms remind of mammalian *in vitro* immune responses that are separated from the architecture of the lymphoid organs. The next big step in the evolution of the immune system is the emergence of highly sophisticated SLTs dedicated to the formation of GCs. This lead to a drastic increase in Ab-affinities, whereas T cell responses have not changed much during evolution of Gnathostomata. On the other hand, even in the absence of such dedicated lymphoid structures, T cells have consistently shown the resourcefulness to initiate immune responses in order to protect the host.

Resourceful Immunity and the Conversion of Unconventional Lymphoid Organs

We discussed above how local inflammation can lead to the emergence of so-called TLTs. A *de novo* appearance of proliferating lymphocyte aggregations defined as inducible bronchus-associated lymphoid tissues (iBALTs) was observed in the lungs of $LT\alpha^{-/-}$ mice, which most likely served as the priming site for virus-specific CTLs [118]. The iBALT structures have further been demonstrated to be sufficient for the maintenance of immunological memory, therewith also providing the structural requirements of secondary immune responses [114]. iBALTs are induced after airway-infection or inflammation and are organized by B cell follicles around networks of FDCs, separated by interfollicular regions containing DCs and T cells, with high endothelial venules (HEV) expressing the L-selectin ligand peripheral node addressin (PNAd), facilitating the recruitment of naïve T cells [118, 194]. These highly organized lymphoid structures potentially facilitate T- and B cell activation with class switch as described earlier.

The list of organs capable of developing TLTs upon chronic inflammation is a long one. Chronic inflammatory diseases in the CNS as well as in other organs such as the joints or thyroid develop features reminiscent of lymphoid tissues [195-199]. Several reports have described how prolonged inflammation during MS can lead to the formation of B cell follicles including FDCs in the meninges of the CNS [200, 201]. Similarly, upon cryptogenic fibrosing alveolitis, the lung will form tissues with ectopic B cell follicles [202, 203]. Chronic inflammatory diseases of the gut such as ulcerative colitis or Crohn`s disease cause the neogenesis of gut-associated lymphoid tissue (GALT) [204-206] and chronic hepatitis C patients develop FDC-containing ectopic follicles in the liver [207, 208]. Finally, T cell aggregates with HEV-like vessels can be found in livers of patients with primary sclerosing cholangitis [209, 210]. Thus, lymphoid neo-organogenesis can be found within many organs in patients with a variety in chronic inflammatory diseases of infectious or autoimmune origin.

However, as can be seen in the many examples described above, the site of lymphoid neo-organogenesis usually seems to be connected to the site of inflammation. Surprisingly, we recently discovered that upon s.c. delivery of antigen, proliferating lymphocyte aggregates were found in the periportal areas of the liver [104], meaning that the site of inflammation was distant from the site of immune cell priming. This implies that antigen-sampling DCs coming from the skin actively selected the liver as a priming site for T cells. In fact, we showed that APCs including DCs picked up microspheres from s.c. immunizations and transported them actively into the liver indicating that not only T cells, but also APCs display plasticity and resourcefulness in order to initiate immune responses outside of SLTs. Other groups have also demonstrated that the liver can facilitate local T cell priming [211]. The liver therefore distinguishes itself from other organs developing TLTs in that it can prime T cell responses both locally and distally to the site of antigen exposure. Interestingly, upon liver transplantation in patients, recipients often inherit allergies from their organ-donors [212], indicating that the liver is a common host to a variety of cells of the adaptive immune response. Just as thymus and BM are places of adult hematopoiesis, the liver carries out this task during embryogenesis and therefore serves transiently as a lymphoid organ [213]. Perhaps the liver "remembers" its task in adulthood and can resume its original function if needed. However, the lymphoid aggregates found in the liver, that were described to raise CMI in the absence of SLTs, did not show any evidence of providing the structural elements needed for B cell priming, such as FDCs or peanut agglutinin (PNA) positive markers [104]. CMI has also been shown to be initiated in other organs such as BM. For example, treatment of mice with the Mel-14 monoclonal Ab to the LN homing-receptor CD62L, limits trafficking of naïve and central memory T cells into the LN through HEVs [214, 215]. Upon intranasal influenza infection of Mel-14-treated splenectomized mice, virus-specific T cells were found in the BM, suggesting that BM can supplant the SLT as a site of primary immune response [157]. The possibility to find effector T cells in the BM due to contamination from the blood or ongoing virus infection within the BM was excluded in these studies. Finally, a study has described that blood-borne antigen can lead to proliferation and differentiation of naïve T cells into effector and memory cells within the BM which is driven by BM-resident DCs [117].

SLTs are a convenient place to concentrate Ag and T cells and in their meeting within SLTs speed up the process of T cell priming. Nonetheless, I have highlighted here several evidences, that in the absence of SLTs, the resourcefulness and pressure on immunity is sufficient to permit T cell priming outside of SLTs even in distant organs such as the liver (subcutaneous Ag) or BM (blood-borne Ag). However, priming then occurs at a decreased pace probably due to the low precursor frequency of antigen-specific T cells. In contrast to T cells however, highly organized lymphoid structures seem to be obligatory for sophisticated B cell activation including high-affinity maturation and class switching.

Interestingly, this hypothesis is strongly supported by what we know about the evolution of immunity. B- and T cell receptors evolved with the conversion of RAG to mediate somatic recombination in jawed vertebrates. While all jawed fish demonstrate adaptive immune functions, Ig class switching appeared from the tetrapod stage of evolution onwards, whereas GC-formation and development of LNs is only visible from birds onwards. The finding that LNs are not a prerequisite for T cell priming has major implications for our understanding of the development and evolution of adaptive immune responses and is of great interest in the context of vaccination. While intranodal immunization has been considered as an effective route of vaccination, recent findings suggest that in the context of CMI, this concept needs to be revised. T cells may not need to be primed in tissue-draining LNs, but they can be primed within the infected/inflamed tissue or even in organs distant to the target tissue where a minimum, yet sufficient, amount of as yet unidentified elements exist to ensure antigen-presentation and sustained proliferation. The liver has been demonstrated to host such an evolutionarily conserved pathway for successful T cell priming independent of lymphoid tissues. This is a good example in which an important function such as host-defense is not dependent on a single pathway, but ensures that fail-safe solutions are in place. The evolutionary conservation of such default pathways is an indication of the strength of the selection pressure on the immune system to find a niche for lymphocyte priming to protect the host.

DISCUSSION

THE ROLE OF NIK-SIGNALING IN CELL-MEDIATED IMMUNITY

It is widely held that non-canonical NFκB-signaling is predominately involved in the formation of SLTs. Consequently, the apparent immunodeficiency of the $NIK^{aly/aly}$ strain was considered as evidence for the requirement of SLTs in the formation of CMI [23]. In recent years, it has however become evident that non-canonical NFκB-signaling is also applied by other cell types such as B cells, osteoclasts, cancer cells and also by DCs and T cells, suggesting a role of non-canonical NFκB-signaling in adaptive immune responses [103-106, 124, 125]. The inability of $NIK^{aly/aly}$ mice to mount CMI is independent of the developmental lymphoreticular malformations but a result of the interrupted NIK-signaling in hematopoietic cells [104]. Yet the mechanistic consequences of lesioned NIK-signaling in DCs and T cells remain poorly understood or might have been wrongly interpreted.

NIK-SIGNALING IN T CELL FUNCTION

Recently, it has been reported that $NIK^{-/-}$ T cells adoptively transferred into $Rag2^{-/-}$ mice fail to develop encephalitogenic properties and to secrete pro-inflammatory cytokines [103]. The authors concluded that $NIK^{-/-}$ T cells are intrinsically defective and that non-canonical NFκB-signaling in T cells is vital for the acquisition of an effector phenotype. This assumption is corroborated by previous reports showing that $NIK^{aly/aly}$ T cells secrete reduced levels of Il-2 and GM-CSF [105]. Indeed, we confirmed a drastic reduction in the secretion of pro-inflammatory cytokines by $NIK^{aly/aly}$ T cells and their failure to homeostatically expand. In addition we found $NIK^{aly/aly}$ T cells to be anergic towards their cognate antigen and to thus fail to acquire pathogenic properties in the context of autoimmune disease. However, our observation that T cell function is impaired as a result of the loss of NIK in hematopoietic accessory cells and more specifically in DCs challenges the concept of a T cell intrinsic defect in $NIK^{aly/aly}$ mice. Two independent experimental setups demonstrated that the T cell defects are the result of a DC intrinsic utilization of NIK: on one hand the presence of $NIK^{aly/aly}$ DCs together with wt T cells *in vivo* is sufficient to significantly diminish EAE development (Fig. 22) and on the other hand $NIK^{-/-}$ T cells can differentiate into fully functional, auto-aggressive T cells, when NIK is restored in accessory cells (Fig. 19). When the expression of NIK is transgenically restricted to DCs via the CD11c promoter, CMI is fully restored even if T cells are $NIK^{-/-}$ (Fig. 23). One caveat is that CD11c-cre is also active in a small population of T cells. This small population of transgenic NIK-expressing T cells could potentially be involved

DISCUSSION

in the rescue of immune function in DC^{NIK}-$NIK^{-/-}$ BMCs. However, this is most unlikely because i) we did not observe a preferential accumulation of NIK-expressing GFP^+ T cells in the inflamed CNS at peak disease in DC^{NIK}-$NIK^{-/-}$ BMCs (data not shown), ii) T cell function of mixed $Rag1^{-/-}$ +$NIK^{aly/aly}$→$Rag1^{-/-}$ BMCs is fully restored even though the entire T cell compartment is NIK-deficient, and most importantly iii) T^{NIK}-$NIK^{-/-}$ BMCs were not susceptible to EAE.

Interestingly, adoptively transferred adult $NIK^{aly/aly}$ T cells fail to acquire pathogenicity but rather displayed an anergy-like state, although they were primed by NIK-sufficient accessory cells (Fig. 18). Only upon undergoing thymic development in the presence of NIK-sufficient accessory cells could T cells give rise to pathogenic effector cells (Fig. 19). Taken together, this suggests that against current belief, NIK is largely dispensable within T cells. In contrast we suggest that the anergic T cell phenotype observed in $NIK^{aly/aly}$ and $NIK^{-/-}$ mice is caused by defective T cell development due to dysfunctional thymic DCs. We therewith propose that NIK-signaling is critical in DCs in order to licence T cells during thymic development and avoid anergy.

NIK-SIGNALING IN DC FUNCTION

A role for non-canonical NFκB-signaling in DCs has already been suggested, but the precise impact on T cell function has not been addressed before. While on one hand it was claimed that CD40-mediated activation of DCs is independent on NIK [102, 123, 216], other reports have demonstrated that peripheral $NIK^{aly/aly}$ DCs express fewer co-stimulatory molecules and show reduced APC capacity, which then results in diminished T cell proliferation [124]. Furthermore, $NIK^{aly/aly}$ DCs show a decreased ability to induce the expansion of $CD25^+$ $CD4^+$ T cells *in vitro* [124]. Additional reports claim that $NIK^{aly/aly}$ DCs are unable to cross-prime $CD8^+$ T cells to exogenous antigen, involving multiple defects in antigen-processing pathways [125]. Also the secretion of the pro-inflammatory cytokine Il-12 but not Il-6 upon CD40-mediated activation was reported to be decreased in $NIK^{aly/aly}$ DCs [143].

We confirmed that splenic $NIK^{aly/aly}$ DCs express less MHCII and co-stimulatory cytokines (data not shown) and produce less Il-12/Il-23p40 protein. We could further observe a reduction in transcript levels for *Il-12p35* and *Il-23p19* as well as reduced Il-6 protein levels. Furthermore, we now show evidence, that $NIK^{aly/aly}$ DCs are hampered in the priming of $CD4^+$ auto-reactive T cells *in vivo* (Fig. 22).

However, the relevance of NIK-signaling in thymic DCs has until now not been addressed, most likely due to the incomplete understanding of the function of thymic DCs in general.

NIK-Signaling in T cell Development

T cell development requires NFκB-signaling (reviewed in [60]). In particular, thymocytes require predominantly canonical NFκB-signaling at various stages during development. NFκB1 is activated by the pre-TCR, acts pro-apoptotic in negative selection and anti-apoptotic in positive selection, promotes long-term survival and is required to generate T_{regs}. In contrast, non-canonical NFκB-signaling plays a major role in the development of lymphoid structures. Disrupted NIK-signaling leads to the absence of LNs and to structural malformations in thymus and spleen. The structure of the thymus with its well organized cortical and medullary regions is believed to be critical to allow proper T cell development. Within this process, thymocytes are migrating through these specific regions in a highly organized manner, passing different checkpoints where they undergo various selections in order to become fully mature T cells without autoimmune potential. In this context, autoimmune regulator (Aire), a transcription factor mainly expressed by mTECs that is also regulated by the non-canonical NFκB-axis, seems to play an important role in negative selection by driving the expression of peripheral-tissue antigens (PTAs) [217].

However, we have now shown, that not only CMI can be initiated independent on a certain structure of SLTs, but also the tight structure of the thymic stroma is dispensable in order to develop mature T cells: $NIK^{aly/+} \rightarrow NIK^{aly/aly}$ BMCs generated T cells, that underwent thymic development within a disorganized thymus of $NIK^{aly/aly}$ stromal origin. These T cells were able to develop full effector capabilities, showing that the tight stromal structure of the thymus was not critical for T cell development [104]. In contrast, we found that the proper thymic stromal structure of $NIK^{aly/+}$ mice was not sufficient to rescue T cell functionality of $NIK^{aly/aly}$ origin. However, the expression of NIK exclusively in DCs was able to restore functionality within NIK-deficient T cells, although only when DCs were NIK-sufficient during the development of NIK-deficient T cells. This implies that a population of DCs in the thymus is more vital to the development of thymocytes than the thymic stromal structure. Indeed, different populations of thymic DCs have been described within the last years, although their function in T cell development remains elusive. Currently, three phenotypically distinct thymic DC subsets have been described, namely pDCs, $CD172\alpha^{+}CD11b^{+}CD8\alpha^{-/lo}$ migratory cDCs and $CD172\alpha^{-}CD11b^{-}CD8\alpha^{hi}$ thymic resident cDCs [127]. One important function of thymic DCs appears to be negative selection at the CD4 SP stage in T cell development [121, 218-220]. In this context, death due to high-affinity interaction or due to no interaction (death by neglect) are not the only fates of developing thymocytes, but also the generation of nT_{regs} or induction of anergic T cells [221-223]. The mechanistic underpinnings of these processes are up to today poorly understood. Although the complete lack of DCs seems to have little effect on T cell development and negative selection [224], a diverted function of thymic DCs is expected to have consequences on T cell development.

The expression of various CCRs and chemokines in thymic DC subsets has been analyzed, described and compared excessively to expression profiles of splenic DCs, but the physiological relevance for most of them remains unclear [128]. We also profiled thymic DCs of $NIK^{aly/aly}$ mice and found strongly reduced gene expression of *CCR2*, *CCR5*, *CCR6* and *CCR7*. It has been proposed that many of these CCRs are important for the localization and migration of DCs into lymphoid tissues in general including the thymus [220]. We further observed a strong reduction in the expression of the chemokines *CCL17*, *CCL19* and *CCL21* in thymic DCs of $NIK^{aly/aly}$ mice (Fig. 25). Especially *CCL17*, *CCL19* and *CCL21* have been described as important chemokines to direct intrathymic migration of positively selected thymocytes from the cortex to the medulla [128, 225]. Thymic dendritic cells have been proposed to induce FoxP3-expression and the formation of nT_{regs} [226]. Taking this into consideration, our data suggests that $NIK^{aly/aly}$ thymic DCs might be defective in the induction of nT_{regs} (Fig. 27). Recently it became evident, that other T effector cell lineages, such as Th17 cells, can at least partially be licensed already during thymic development [131]. We found that also Tbet and RORγt-expressing $CD4^+$ SP T cells are decreased in thymi of $NIK^{aly/aly}$ mice. (Fig. 27). This phenotype and also the decrease in nT_{regs} is rescued in $NIK^{-/-}$ $CD4^+$ SP T cells of DC^{NIK}→wt BMCs. We therefore propose that NIK-signaling in thymic DCs is crucial to imprint developing T cells to subsequently aquire full effector capabilities and to avoid progression into an anergic state.

DOES NIK-SIGNALING EQUAL NON-CANONICAL NFκB-SIGNALING?

Several reports claimed that under specific circumstances, NIK signals into the canonical NFκB-pathway [67, 81, 82].

Curiously, even though both $NFκB2^{-/-}$ and $NIK^{aly/aly}$ mice show impaired T and B cell responses, both have also been reported to contain lymphocyte infiltration into various organs similar to that of $Aire^{-/-}$ mice [227-231]. However, the autoimmune phenotype in $NFκB2^{-/-}$ and $NIK^{aly/aly}$ mice seem to originate from the stromal compartment, as transplantation of $NFκB2^{-/-}$ and $NIK^{aly/aly}$ thymi was sufficient to induce the breakdown in self-tolerance [227, 228]. The phenotype of tolerance breakdown in both $NFκB2^{-/-}$ and $NIK^{aly/aly}$ mice seems to be mediated by the absence of Aire-signaling in mTECs [227] via the non-canonical NFκB-signaling cascade [228]. Instead, the impairment of T cell responses in $NIK^{aly/aly}$ mice results from disrupted NIK-signaling in hematopoetic cells. Surprisingly, in contrast to NIK$^{aly/aly}$ mice, NFκB2$^{-/-}$ mice do not show any defect in T$_{reg}$ generation neither in spleen nor in thymi [227]. Furthermore, DCs of NFκB2$^{-/-}$ mice express increased levels of MHCII and co-stimulatory molecules as well as an enhanced ability to induce CD4^{+} T cell responses [232]. We also observed that RelB$^{-/-}$→wt BMCs are fully susceptible to EAE (Figure 29). These observations suggest, that the signaling-cascade, by which NIK executes its function in CMI, is not via the p52-dependent non-canonical NFκB-pathway. However, by which mechanism and signaling cascade NIK conducts its function in thymic DCs in order to licence developing T cells remains unclear.

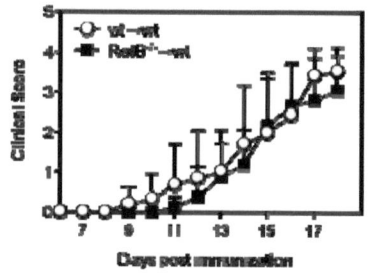

Figure 29: RelB$^{-/-}$→wt BMCs are susceptible to EAE. RelB$^{-/-}$→wt and wt→wt BMCs were immunized with MOG$_{35-55}$/CFA and observed for clinical signs of EAE. Shown is one preliminary experiment (n=4-6).

NFκB-SIGNALING IN AUTOIMMUNITY

NFκB is constitutively activated in many autoimmune diseases, including diabetes type 2, systemic lupus erythematosus and rheumatoid arthritis [84]. Because NFκB plays a central role in differentiation, activation and survival of cells, it contributes to autoimmune diseases in multiple ways. Because NFκB acts in both T and B cell development, defects in these processes may initiate pathogenicity. NFκB plays a pro-apoptotic role in the negative selection of self-reactive B and T cells. In the periphery, NFκB is essential for the maintenance of B cell homeostasis. Defects may lead to prolonged survival of B cells with autoimmune reactivity [233]. NFκB is further required for the development of mTECs and the organization of the thymic stroma, and the development and differentiation of DCs [228, 234-236]. Mice lacking mTECs, thymic structure or DC subtypes are often characterized by severe autoimmunity with autoreactive T cells, multiple organ lymphocytic infiltrates and in some cases early mortality [228, 235]. NFκB2-deficiency leads to a milder phenotype, likely because of compensation by NFκB1. In the absence of NFκB2, NFκB1 can form heterodimers with RelB and might therewith replace p52/RelB. Combined Bcl-3/NFκB2-deficiency (complete lack of NFκB-signaling) leads to a dramatic autoimmune phenotype, caused by a complete lack of mTECs and consequent loss of negative selection of auto-reactive T cells [237]. Also defects in the upstream activators of the canonical and non-canonical NFκB-pathways effect the establishment of self-tolerance. For example TRAF6-deficiency also leads to thymic disorganization, reduced T_{reg}-production, absence of mature mTECs and autoimmunity [236, 238]. Although TRAF6 classically activates the canonical NFκB-pathway, it might also indirectly effect the non-canonical pathway as TRAF6-deficiency leads to the lack of RelB expression in mTECs, that again leads to reduced T_{reg} generation [238]. In contrast, the lack of TRAF3, which is a negative regulator on the non-canonical NFκB-pathway, leads to an increased activation of NFκB2. The B cell-specific TRAF3$^{-/-}$ results in prolonged B cell survival, greatly an expanded B cell compartment and autoimmune reactivity [233].

Many autoimmune diseases are characterized by the development of TLTs at the site of inflammation (as described above). The non-canonical NFκB-pathway may be implicated in the generation of TLTs as the constitutive expression of LTβ has been shown to induce TLTs [239]. Furthermore, most autoimmune diseases are displaying elevated levels of the cytokine BAFF that is needed for B cell maturation and survival [240] and signals via the non-canonical pathway. Additionally, NIK-deficient mice are not only resistant to EAE, but also largely resistant to rheumatoid arthritis and inflammation-induced osteoclastogenesis [241].

Lastly, most autoimmune diseases are characterized by the expression of many pro-inflammatory cytokines, such as Il-6, Il-1, TNF-α and Il-17, that play an important role in sustaining the inflammation by recruiting and activating immune cells to the site of inflammation. Most of these pro-inflammatory cytokines induce NFκB-pathways and treatments targeting the signaling via these cytokines seem to be effective in attenuating pathogenicity [242].

Taken together, improperly regulated NFκB-signaling leading to its constitutive activation results in autoimmunity. Autoimmune diseases may be caused by malfunctioning lymphocytes whose apoptotic pathways are blocked by abnormal NFκB-activation, enabling the survival of self-reacting cells. The multiple roles of NFκB in autoimmune diseases make it an important pharmaceutical target. However, because NFκB is involved in many crucial roles that maintain health in general, the challenge lies in finding a balance between avoiding deleterious side effects caused by systemic NFκB-inhibition and the treatment of disease-specific symptoms.

CONCLUSION

Alymphoplastic NIK$^{aly/aly}$ mice lack LN and PPs due to a point mutation in NIK, which is the main mediator of non-canonical NFκB-signaling. They are highly immunodeficient and resistant to EAE. Defining the reason of its immunodeficiency, we found that T cell-driven immune responses are independent of SLT structures, while B cell responses strongly depend on the topography of dedicated lymphoid tissues. This phenomenon can be explained by the emergence of high affinity maturation and class switch of antibodies during evolution, which coincided with the appearance of defined lymphoid structures, whereas T cells were generally as effective as they are today before LNs developed.

While the EAE resistance of NIK$^{aly/aly}$ mice was not connected to the lack of SLTs, it was caused by the lack of NIK in hematopoietic cells. CMI could not be induced in mice that were deficient of NIK in all immune cells but had normal SLTs. However, NIK was not required in T cells themselves, but was rather essential in thymic DCs to allow proper T cell development. We have therewith not only demystified the true reason of CMI immunodeficiency in NIK$^{aly/aly}$ mice but also identified a novel function in thymic DCs: the capacity to shape the T effector repertoire early during development.

Discussion

EXPERIMENTAL PROCEDURES

Mice, bone marrow transplantation and splenectomy

C57BL/6 and $Rag1^{-/-}$ mice were purchased from Janvier Laboratories, France. *Alymphoplasia* mice (*Map3k14^{aly}* mice here depicted as $NIK^{aly/aly}$) were obtained from Clea laboratories Japan and bred in house under SPF-conditions. Heterozygous $NIK^{aly/+}$ mice were used as controls for homozygous $NIK^{aly/aly}$ mice. $NIK^{-/-}$ mice on 129Sv/Ev background were generously provided by Robert D. Schreiber (Washington University, St. Louis, MO, USA) and bred onto C57BL/6 background in house for ten generations. Both $NIK^{aly/aly}$ and $NIK^{-/-}$ mice were maintained by heterozygous breedings. 2d2 (MOG-TCR Tg) mice were provided by V. Kuchroo (Harvard Medical School, Boston, Massachusetts), $LT\beta R^{-/-}$ and $LT\alpha^{-/-}$ mice were provided by A. Aguzzi and M. Heikenwalder (University Hospital Zurich, Zurich, Switzerland) and OTII and OTI mice were purchased from Jackson Laboratories. Luciferase (pbActin-Luciferase) transgenic mice were obtained from C. Contag (UCSF) and crossed to the 2D2 mice (Luc-2D2). *Plt/plt* mice were obtained from B. Ludewig (Kantonsspital St.Gallen, Switzerland). *CD11cDTR* mice were provided by Steffen Jung (Weizmann Institute of Science, Rehovot, Israel). *CD11cCre* mice were generously provided by Boris Reizis (Columbia University, New York, NY, USA). $R26Stop^{FL}NIK^{wt}$ and all other mice were maintained under specific pathogen-free conditions.

Bone marrow chimeric mice (BMCs) were generated as described previously [243, 244]. Briefly, mice were lethally irradiated with a split-dose of 1100 rad. Femur, tibia and pelvis of donor animals were flushed with PBS to obtain BM stem cells. 10×10^6 cells were injected intravenously (i.v.) per mouse. Mice were treated with 0.2% BORGAL in drinking water for 3 weeks to prevent bacterial infections. Mice were splenectomized as described previously [245]. Animal experiments were approved by the Swiss veterinary Office (68/2003, 70/2003, 10/2006, 13/2006 and 55/2009; Zurich, Switzerland).

Induction of EAE and DTx treatment

MOG_{35-55} peptide (MEVGWYRSPFSRVVHLYRNGK) was obtained from GenScript. EAE was induced as described previously [246]. Briefly, mice were immunized s.c. with 200µg of MOG_{35-55} emulsified in CFA (Difco) and two i.p. injections of 200ng pertussis toxin on day 0 and 2. BMCs did not receive pertussis toxin. For adoptive transfer MOG-reactive lymphocytes were generated as described [246]. For EAE experiments with DTx treatment, mice were injected i.p. with 400 ng DTx (Calbiochem) one day before immunization and then 200 ng DTx every second day for the entire length of the experiment. Each time point shown is the average disease score of each group ± SEM.

Leukocyte isolation

Mice were euthanized with CO_2 and various organs were removed to isolate leukocytes: For isolating lung cells, lungs were incubated with DNase (0.5mg) / Liberase (1mg/ml) (Roche) for 30 min at 37° C. Spleen, LNs, thymus and lung were homogenized and bone marrow cells were isolated by flushing the bones with PBS. Cells were strained through a 100 μm nylon filter (Fisher) and washed. Erythrocytes of whole blood, BM and spleen were lysed. For isolating hepatic non-parenchymal cells, the liver was incubated with DNase/ Liberase for 30 min at 37° C, homogenized and centrifuged at RT for 2 min at 50G. The supernatant was then centrifuged at 1500 rpm for 10 min and the pellet was resuspended in 30% Percoll (Pharmacia) and centrifuged at 12000 rpm for 30 min at 4° C. The interphase cells were collected and washed. For isolating intestinal lymphocytes, intestines were opened longitudinally, washed and cut into small pieces. Tissues were then incubated with DNase/Liberase and leukocytes were isolated using a percoll gradient as described above. Isolation of CNS lymphocytes has been described previously [245].

Proliferation assay

Mice were injected i.v. with 20×10^6 CFSE (carbofluorescein diacetate succinimidyl ester)-labeled (Invitrogen-Molecular Probes) (10μM) splenocytes obtained from either 2D2, OT-II or OT-I TcR Tg mice or with 8×10^6 CFSE-labeled naïve $CD4^+$ 2D2 Tg T cells (isolated with $CD4^+CD62L^+$ isolation kit from Miltenyi). Mice were subsequently immunized s.c. with 200 μg of MOG_{35-55} /CFA (Adjuvant complete H37 Ra, DIFCO) (for 2D2), $OVA_{323-339}$ /CFA (for OT-II) or with a 1:1 mix of irradiated 2×10^6 B16.F10-GM-CSF/B16.F10-OVA cells (for OT-I). 4 or 5 dpi (12 dpi for OT-I) mice were sacrificed and spleen, LNs (if present) and livers were analyzed by FACS for the proliferation of $CD4^+$ T cells using the clonotypic TcR and CFSE-fluorescence (2D2: TCR Vα3.2 Ab, OT-II and OT1: Vα2 Ab).

Histology

Tissues were freshly snap-frozen in liquid nitrogen. To determine infiltration of inflammatory cells, tissue sections were stained with hematoxylin and eosin (H&E) or with the following mouse-specific antibodies (Abs) as previously described [246]: anti-CD11c (Jackson ImmunoResearch

Labs), anti-CD11b (BMA Biomedicals), anti-CD3, anti-CD4, anti-FDC M1 and anti-Thy1.1 (BD-Pharmingen). GC cells were stained with peanut agglutinin (PNA; Vector Laboratories).

Flow cytometry

For cell surface staining, we used the following antibodies (Abs): CD11c, CD11b, I^{ab}, CD80, CD86, CD172a, CD45RA, CD4, CD8, Il-12/Il-23p40, Vα3.2, Vβ11, Thy1.1 (BD Pharmingen). Intracellular FoxP3 stain was performed according to manufacturers instructions (eBioscience). Cells were incubated with Abs at the optimal concentration for 20 min at 4° C and cells were analyzed with FACS Canto II (BD Pharmingen) and FACS Diva Software. Postaquisaition analysis was performed with either FACS Diva or FlowJo (TreeStar) software. Cytofluorometric analysis of CNS invading lymphocytes has been described previously [247]. For intracellular cytokine staining cells were treated with GolgiPlug (BD Biosciences) for the last 4 hours of culture. After Ab surface staining, cells were permeabilized with Cytofix/Cytoperm (BD Biosciences) according to manufacturer's recommendations and stained intracellular with Il-12/Il-23p40-specific Ab (BD Pharmingen).

To trace the distribution of Ag after immunization, mice were injected s.c. with 200 μl Fluoresbrite^{TM} Carboxylate (YG) or Polychromatic Red (PR) 1.0 micron microspheres (Polysciences) emulsified in CFA. 7 days after injection mice were euthanized with CO_2 and organs were removed to isolate lymphocytes as described above. Single cell suspensions were analyzed by FACS for the presence of FITC$^+$ or PE$^+$ cells. For FITC skin painting, mice were painted on the shaved flanks with 100 μL of 5 mg/ml fluorescein isothiocyanate (FITC) (Molecular Probes) dissolved in 1:1 acetone:dibutylphtalate. On day 1, mice were euthanized with CO_2 and organs removed and analyzed by FACS as described above.

Delayed-type hypersensitivity (DTH) assay

Mice were immunized s.c. with 100 μg/flank of MOG$_{35-55}$ peptide , or KLH (Sigma) emulsified in CFA. 11dpi, mice were challenged by injecting 10 μg/10μl of MOG$_{35-55}$ peptide, KLH, PBS into the dorsal surface of the ear. DTH responses were determined by measuring the ear thickness using a calliper micrometer (Mitutoyo) 24 h after challenge and Δ ear swelling was established by the increase in ear thickness over baseline (pre-challenge ear thickness).

Enzyme-linked immunosorbent assay (ELISA) for antibody detection

Plates were coated with 10μg rMOG$_{1-121}$ in 0.1 M NaHCO$_3$ (pH 9.6) at 4° C overnight or KLH (Sigma) and blocked with 1% (w/v) bovine serum albumin (BSA). Diluted sera were incubated for 2h at RT. After washing, peroxidase-conjugated antibodies to mouse Ig's, IgG, IgA, IgM (Sigma) were added (1:1000 diluted) and incubated for 1h at RT. Plates were washed and chromogen (Biosource) was added. Absorbance was measured on a micro plate reader (450 nm) (Bio-Rad).

Enzyme-linked immunospot analysis (Elispot)

2 x 10^5 cells were plated in medium containing 10% FCS and MOG$_{35-55}$ 50 μg/ml in 96 well plates (Millipore) coated with the capture Ab against either IFNγ or Il-17A[36]. Elispots were revealed as described previously [36] and subsequently analyzed on an Elispot reader (CTL immunospot).

Bioluminescence imaging

To visualize Luc-2D2-cells, mice were injected i.p. with 3 mg of Luciferin (Xenogen) prior to bioluminescence imaging using an IVIS100 imaging station (Xenogen). The luminescent image was overlaid on the photographic image.

Bromodeoxyuridine (BrdU) treatment

Mice were immunized s.c. with MOG$_{35-55}$/CFA. 7 dpi BrdU (BD Pharmingen) (2.5 mg) was injected i.p. 30 min before the mice were sacrificed and analyzed for proliferating (BrdU$^+$) CD4$^+$ T cells by flow cytometry with anti-BrdU Ab (eBioscience).

Tumor induction

Mice were s.c. vaccinated into one flank with irradiated (6000 rads) 1x10^6 B16.F10-GM-CSF cells. At day 12 after vaccination, mice were injected with live 2x10^5 B16.F10-Luc cells into the opposite flank. Each time point shown is the average tumor size of each group ± SEM measured using a caliper.

Isolation of splenic DCs and *in vitro* stimulation

Spleens were removed under sterile conditions. Each spleen was injected with a cocktail of 1 mg/ml liberase and 0.5 mg/ml DNaseI (Roche) in medium and incubated at 37° C for 20 min. Single cell suspensions were prepared by homogenizing the tissue between glass slides and filtering through 70 μm cell strainers followed by erythrocyte lysis. Splenic DCs were isolated with CD11c$^+$ positive magnetic selection according to the manufacturer's instructions (Miltenyi Biotech).

DCs were plated at a concentration of 1×10^6 cells/ml in RPMI-1640 medium supplemented with 10% FCS, 1% L-glutamine and 1% penicillin-streptavidin (Invitrogen Life Technologies), and stimulated for 6 to 24 hours at 37° C 5% CO_2 with 10 µg/ml α-CD40 (BioXCell; FGK 4.5) and 20 ng/ml IFNγ (Peprotec). Supernatants were analyzed for Il-6 and Il-12/Il-23p40 using ELISA according to manufacturer's instructions (BD Pharmingen).

Isolation of thymic DCs and *in vitro* co-culture assays

Thymic DCs were isolated as previously described [248]. Briefly, thymi were digested in IMDM containing 2% FCS, 25mM HEPES, 0.4mg/ml Collagenase D (Roche) and DNAse for 40 min at 37° C. Afterwards, high-density cells were separated from low-density cells by using a discontinous Percoll density gradient (GE Healthcare, ρ=1.115 and ρ1.055). After removal from the gradient cells were washed and stained with antibodies against CD11c, CD8, CD172a and CD45RA, followed by sorting into migratory DCs (CD11c$^+$ CD172a$^+$), resident DCs (CD11c$^+$ CD172a$^-$) and pDCs (CD11c^{int} CD45RA$^+$) on a FACSAria (BD). Purity was routinely >95%.

For *in vitro* culture 20.000 DCs were cultured with 100.000 sorted and CFSE-labeled CD4$^+$ thymocytes or peripheral T cells in the presence of 10 ug/ml MOG$_{35-55}$ and 10 ng/ml Il-7. Analysis was performed after 72 hours of culture.

Peripheral CD4$^+$ T cell purification, *in vitro* stimulation and adoptive transfer

Splenocyte single cell suspensions were prepared as described above. CD4$^+$ T cells were purified with CD4$^+$ negative magnetic selection according to the manufacturer's instructions (Miltenyi Biotech). The purity was routinely >95% as confirmed by flow cytometry.

For *in vitro* stimulations of splenic CD4$^+$ T cells, 3×10^6 CD4$^+$ T cells/ml were cultured in RPMI-1640 medium supplemented with 10% FCS, 1% L-glutamine, 1% penicillin-streptavidin. Polyclonal CD4$^+$ T cell activation was performed with 5 µg/ml plate-bound α-CD3 and α-CD28 for 48 hours. For antigen-specific CD4$^+$ T cell activation whole 2d2 splenocytes were stimulated with 20 µg/ml MOG$_{35-55}$ and 5 µg/ml soluble α-CD28 for 48 hours. Supernatants were harvested and concentrations of IFNγ, Il-17, GM-CSF, Il-2 and Il-4 were quantified by ELISA according to manufacturers instructions (BD Pharmingen).

For adoptive transfer experiments, 2×10^6 purified splenic CD4$^+$ T cells in PBS were injected i.v. into *Rag1$^{-/-}$* mice. T cell reconstitution of *Rag1$^{-/-}$* mice was monitored weekly by tail bleeding and flow cytometry.

EXPERIMENTAL PROCEDURES

RNA isolation and quantitative (q) RT-PCR

Total RNA was isolated according to manufacturer's instructions (RNeasy Mini Plus Kit, Qiagen). Reverse transcription was performed using random hexamer primers and Moloney murine leukemia virus reverse transcriptase (splenic DCs) or Superscript II (thymic DCs) (Invitrogen Life Technologies). cDNA was analyzed by quantitative real-time PCR (qRT-PCR) (Biorad) in duplicates using SYBR Green PCR Mastermix (Invitrogen) or hydrolysis TAMRA probes (Roche). The expression level of each gene was normalized to *HPRT* or *DNA Polymerase II*. The following primers purchased from Operon Technologies were used:

HPRT (5'-GACCGGTCCCGTCATGC-3', 5'-TCATAACCTGGTTCATCATCGC-3'), *DNA Polymerase II* (5'-CTGGTCCTTCGAATCCGCATC-3', 5'-GCTCGATACCCTGCAGGGTCA-3'),
Il-12/Il-23p40 (5'-GACCATCACTGTCAAAGAGTTTCTAGAT-3', 5'-AGGAAAGTCTTGTTTTTGAAATTTTTTAA-3'),
Il-12p35 (5'-TACTAGAGAGACTTCTTCCACAACAAGAG-3', 5'-TCTGGTACATCTTCAAGTCCTCATAGA-3'),
Il-23p19 (5'-AGCGGGACATATGAATCTACTAAGAGA-3', 5'-GTCCTAGTAGGGAGGTGTGAAGTTG-3'),
TLR9 (5'-GCCTTCGTGGTGTTCGATAAGG-3', 5'-GAGGTTCTCGAAGAGCGTCTGG-3'),
CCL17 (5'-TACTTCAAAGGGGCCATTCCT-3', 5'-GCCTTGGGTTTTTCACCAATCT-3'),
CCL19 (5'-GGCCTGCCTCAGATTATCTGCCAT-3', 5'-GGAAGGCTTTCACGATGTTCC-3'),
CCL21 (5'-GGACCCAAGGCAGTGATGGAG-3', 5'-CTTCCTCAGGGTTTGCACATAG-3'),
CCR2 (5'-ACAAGCACTTAGACCAGGCCAT-3', 5'-AAACTGGGCACTGTTTGC-3'),
CCR5 (5'-ACTGCTGCCTAAACCCTGTCA-3', 5'-GTTTTCGGAAGAACACTGAGAGATAA-3'),
CCR6 (5'-TCCATCATCATCTCAAGCCCTACA-3', 5'-AGGGGTGAAGAACCCAAAGAACA-3'),
CCR7 (5'-ACCATGGACCCAGGGAAC-3', 5'- GGTATTCTCGCCGATGTAGTCAT-3'),
CCR9 (5'-TGGCTTGTGTTCATTGTGGGCA-3', 5'-ATCCATTGACCAGCAGCAGCAA-3').

REFERENCES

1. Charles A Janeway, P.T., Mark Walport, Mark Shlomchik, *Immunobiology: The Immune Sytem in Health and Disease*. 2004(6th edition).

2. Petrie, H.T. and J.C. Zuniga-Pflucker, *Zoned out: functional mapping of stromal signaling microenvironments in the thymus*. Annu Rev Immunol, 2007. **25**: p. 649-79.

3. Carpenter, A.C. and R. Bosselut, *Decision checkpoints in the thymus*. Nat Immunol, 2010. **11**(8): p. 666-73.

4. Merkenschlager, M., et al., *How many thymocytes audition for selection?* J Exp Med, 1997. **186**(7): p. 1149-58.

5. Starr, T.K., S.C. Jameson, and K.A. Hogquist, *Positive and negative selection of T cells*. Annu Rev Immunol, 2003. **21**: p. 139-76.

6. Takahama, Y., *Journey through the thymus: stromal guides for T-cell development and selection*. Nat Rev Immunol, 2006. **6**(2): p. 127-35.

7. Pabst, R., *Plasticity and heterogeneity of lymphoid organs. What are the criteria to call a lymphoid organ primary, secondary or tertiary?* Immunol Lett, 2007. **112**(1): p. 1-8.

8. Fu, Y.X. and D.D. Chaplin, *Development and maturation of secondary lymphoid tissues*. Annu Rev Immunol, 1999. **17**: p. 399-433.

9. Mach, J., et al., *Development of intestinal M cells*. Immunol Rev, 2005. **206**: p. 177-89.

10. Aloisi, F. and R. Pujol-Borrell, *Lymphoid neogenesis in chronic inflammatory diseases*. Nat Rev Immunol, 2006. **6**(3): p. 205-17.

11. Gutcher, I. and B. Becher, *APC-derived cytokines and T cell polarization in autoimmune inflammation*. J Clin Invest, 2007. **117**(5): p. 1119-27.

12. Mosmann, T.R., et al., *Two types of murine helper T cell clone. I. Definition according to profiles of lymphokine activities and secreted proteins*. J Immunol, 1986. **136**(7): p. 2348-57.

13. Palmer, M.T. and C.T. Weaver, *Autoimmunity: increasing suspects in the CD4+ T cell lineup*. Nat Immunol, 2010. **11**(1): p. 36-40.

14. O'Shea, J.J. and W.E. Paul, *Mechanisms underlying lineage commitment and plasticity of helper CD4+ T cells*. Science, 2010. **327**(5969): p. 1098-102.

15. Groux, H., et al., *A CD4+ T-cell subset inhibits antigen-specific T-cell responses and prevents colitis*. Nature, 1997. **389**(6652): p. 737-42.

16. Nurieva, R.I., et al., *Generation of T follicular helper cells is mediated by interleukin-21 but independent of T helper 1, 2, or 17 cell lineages*. Immunity, 2008. **29**(1): p. 138-49.

17. Steinman, L., *A brief history of T(H)17, the first major revision in the T(H)1/T(H)2 hypothesis of T cell-mediated tissue damage*. Nat Med, 2007. **13**(2): p. 139-45.

18. Marks, B.R. and J. Craft, *Barrier immunity and IL-17*. Semin Immunol, 2009. **21**(3): p. 164-71.

19. Veldhoen, M., et al., *Transforming growth factor-beta 'reprograms' the differentiation of T helper 2 cells and promotes an interleukin 9-producing subset*. Nat Immunol, 2008. **9**(12): p. 1341-6.

20. Dardalhon, V., et al., *IL-4 inhibits TGF-beta-induced Foxp3+ T cells and, together with TGF-beta, generates IL-9+ IL-10+ Foxp3(-) effector T cells*. Nat Immunol, 2008. **9**(12): p. 1347-55.

21. Lee, Y.K., et al., *Late developmental plasticity in the T helper 17 lineage.* Immunity, 2009. **30**(1): p. 92-107.

22. Ghoreschi, K., et al., *Generation of pathogenic T(H)17 cells in the absence of TGF-beta signalling.* Nature, 2010. **467**(7318): p. 967-71.

23. Hofmann, J., et al., *B-cells need a proper house, whereas T-cells are happy in a cave: the dependence of lymphocytes on secondary lymphoid tissues during evolution.* Trends Immunol, 2010.

24. Rioux, J.D. and A.K. Abbas, *Paths to understanding the genetic basis of autoimmune disease.* Nature, 2005. **435**(7042): p. 584-9.

25. O'Shea, J.J., A. Ma, and P. Lipsky, *Cytokines and autoimmunity.* Nat Rev Immunol, 2002. **2**(1): p. 37-45.

26. Compston, A. and A. Coles, *Multiple sclerosis.* Lancet, 2002. **359**(9313): p. 1221-31.

27. McFarland, H.F. and R. Martin, *Multiple sclerosis: a complicated picture of autoimmunity.* Nat Immunol, 2007. **8**(9): p. 913-9.

28. Jones, J.L. and A.J. Coles, *New treatment strategies in multiple sclerosis.* Exp Neurol, 2010. **225**(1): p. 34-9.

29. Kurtzke, J.F., *A reassessment of the distribution of multiple sclerosis.* Acta Neurol Scand, 1975. **51**(2): p. 137-57.

30. Kurtzke, J.F., *A reassessment of the distribution of multiple sclerosis. Part one.* Acta Neurol Scand, 1975. **51**(2): p. 110-36.

31. Ramagopalan, S.V., et al., *Multiple sclerosis: risk factors, prodromes, and potential causal pathways.* Lancet Neurol, 2010. **9**(7): p. 727-39.

32. Oksenberg, J.R. and S.L. Hauser, *Genetics of multiple sclerosis.* Neurol Clin, 2005. **23**(1): p. 61-75, vi.

33. De Jager, P.L., et al., *Meta-analysis of genome scans and replication identify CD6, IRF8 and TNFRSF1A as new multiple sclerosis susceptibility loci.* Nat Genet, 2009. **41**(7): p. 776-82.

34. Kurtzke, J.F., G.W. Beebe, and J.E. Norman, Jr., *Epidemiology of multiple sclerosis in U.S. veterans: 1. Race, sex, and geographic distribution.* Neurology, 1979. **29**(9 Pt 1): p. 1228-35.

35. Orton, S.M., et al., *Sex ratio of multiple sclerosis in Canada: a longitudinal study.* Lancet Neurol, 2006. **5**(11): p. 932-6.

36. Whitacre, C.C., *Sex differences in autoimmune disease.* Nat Immunol, 2001. **2**(9): p. 777-80.

37. Ascherio, A. and K.L. Munger, *Environmental risk factors for multiple sclerosis. Part I: the role of infection.* Ann Neurol, 2007. **61**(4): p. 288-99.

38. Levin, L.I., et al., *Temporal relationship between elevation of epstein-barr virus antibody titers and initial onset of neurological symptoms in multiple sclerosis.* JAMA, 2005. **293**(20): p. 2496-500.

39. Johnson, K.P., *Glatiramer acetate and the glatiramoid class of immunomodulator drugs in multiple sclerosis: an update.* Expert Opin Drug Metab Toxicol, 2010. **6**(5): p. 643-60.

40. Revel, M., *Interferon-beta in the treatment of relapsing-remitting multiple sclerosis.* Pharmacol Ther, 2003. **100**(1): p. 49-62.

41. Bielekova, B., et al., *Regulatory CD56(bright) natural killer cells mediate immunomodulatory effects of IL-2Ralpha-targeted therapy (daclizumab) in multiple sclerosis.* Proc Natl Acad Sci U S A, 2006. **103**(15): p. 5941-6.

42. Bar-Or, A., et al., *Rituximab in relapsing-remitting multiple sclerosis: a 72-week, open-label, phase I trial.* Ann Neurol, 2008. **63**(3): p. 395-400.

43. Robak, T., *Alemtuzumab in the treatment of chronic lymphocytic leukemia.* BioDrugs, 2005. **19**(1): p. 9-22.

REFERENCES

44. Watson, C.J., et al., *Alemtuzumab (CAMPATH 1H) induction therapy in cadaveric kidney transplantation-- efficacy and safety at five years.* Am J Transplant, 2005. **5**(6): p. 1347-53.

45. Hale, G. and H. Waldmann, *Recent results using CAMPATH-1 antibodies to control GVHD and graft rejection.* Bone Marrow Transplant, 1996. **17**(3): p. 305-8.

46. Olitsky, P.K. and R.H. Yager, *Experimental disseminated encephalomyelitis in white mice.* J Exp Med, 1949. **90**(3): p. 213-24.

47. Kamradt, T., et al., *Pertussis toxin prevents the induction of peripheral T cell anergy and enhances the T cell response to an encephalitogenic peptide of myelin basic protein.* J Immunol, 1991. **147**(10): p. 3296-302.

48. Shive, C.L., et al., *The enhanced antigen-specific production of cytokines induced by pertussis toxin is due to clonal expansion of T cells and not to altered effector functions of long-term memory cells.* Eur J Immunol, 2000. **30**(8): p. 2422-31.

49. Schreiner, B., F.L. Heppner, and B. Becher, *Modeling multiple sclerosis in laboratory animals.* Semin Immunopathol, 2009. **31**(4): p. 479-95.

50. Huseby, E.S., et al., *A pathogenic role for myelin-specific CD8(+) T cells in a model for multiple sclerosis.* J Exp Med, 2001. **194**(5): p. 669-76.

51. Sun, D., et al., *Myelin antigen-specific CD8+ T cells are encephalitogenic and produce severe disease in C57BL/6 mice.* J Immunol, 2001. **166**(12): p. 7579-87.

52. Dal Canto, M.C. and H.L. Lipton, *Primary demyelination in Theiler's virus infection. An ultrastructural study.* Lab Invest, 1975. **33**(6): p. 626-37.

53. Lipton, H.L., *Theiler's virus infection in mice: an unusual biphasic disease process leading to demyelination.* Infect Immun, 1975. **11**(5): p. 1147-55.

54. Buch, T., et al., *A Cre-inducible diphtheria toxin receptor mediates cell lineage ablation after toxin administration.* Nat Methods, 2005. **2**(6): p. 419-26.

55. Woodruff, R.H. and R.J. Franklin, *Demyelination and remyelination of the caudal cerebellar peduncle of adult rats following stereotaxic injections of lysolecithin, ethidium bromide, and complement/anti-galactocerebroside: a comparative study.* Glia, 1999. **25**(3): p. 216-28.

56. Matsushima, G.K. and P. Morell, *The neurotoxicant, cuprizone, as a model to study demyelination and remyelination in the central nervous system.* Brain Pathol, 2001. **11**(1): p. 107-16.

57. Felts, P.A., et al., *Inflammation and primary demyelination induced by the intraspinal injection of lipopolysaccharide.* Brain, 2005. **128**(Pt 7): p. 1649-66.

58. Hayden, M.S. and S. Ghosh, *Shared principles in NF-kappaB signaling.* Cell, 2008. **132**(3): p. 344-62.

59. Xiao, G., E.W. Harhaj, and S.C. Sun, *NF-kappaB-inducing kinase regulates the processing of NF-kappaB2 p100.* Mol Cell, 2001. **7**(2): p. 401-9.

60. Siebenlist, U., K. Brown, and E. Claudio, *Control of lymphocyte development by nuclear factor-kappaB.* Nat Rev Immunol, 2005. **5**(6): p. 435-45.

61. Ghosh, S. and M. Karin, *Missing pieces in the NF-kappaB puzzle.* Cell, 2002. **109 Suppl**: p. S81-96.

62. Sha, W.C., et al., *Targeted disruption of the p50 subunit of NF-kappa B leads to multifocal defects in immune responses.* Cell, 1995. **80**(2): p. 321-30.

63. Artis, D., et al., *NF-kappa B1 is required for optimal CD4+ Th1 cell development and resistance to Leishmania major.* J Immunol, 2003. **170**(4): p. 1995-2003.

64. Youssef, S. and L. Steinman, *At once harmful and beneficial: the dual properties of NF-kappaB.* Nat Immunol, 2006. **7**(9): p. 901-2.

REFERENCES

65. Darnay, B.G., et al., *Activation of NF-kappaB by RANK requires tumor necrosis factor receptor-associated factor (TRAF) 6 and NF-kappaB-inducing kinase. Identification of a novel TRAF6 interaction motif.* J Biol Chem, 1999. **274**(12): p. 7724-31.

66. Maruyama, T., et al., *Processing of the NF-kappa B2 precursor p100 to p52 is critical for RANKL-induced osteoclast differentiation.* J Bone Miner Res, 2010. **25**(5): p. 1058-67.

67. Ramakrishnan, P., W. Wang, and D. Wallach, *Receptor-specific signaling for both the alternative and the canonical NF-kappaB activation pathways by NF-kappaB-inducing kinase.* Immunity, 2004. **21**(4): p. 477-89.

68. Nadiminty, N., et al., *LIGHT, a member of the TNF superfamily, activates Stat3 mediated by NIK pathway.* Biochem Biophys Res Commun, 2007. **359**(2): p. 379-84.

69. Sanz, A.B., et al., *TWEAK activates the non-canonical NFkappaB pathway in murine renal tubular cells: modulation of CCL21.* PLoS One, 2010. **5**(1): p. e8955.

70. Sanchez-Valdepenas, C., et al., *NF-kappaB-inducing kinase is involved in the activation of the CD28 responsive element through phosphorylation of c-Rel and regulation of its transactivating activity.* J Immunol, 2006. **176**(8): p. 4666-74.

71. Bhattacharyya, S., et al., *Lipopolysaccharide-induced activation of NF-kappaB non-canonical pathway requires BCL10 serine 138 and NIK phosphorylations.* Exp Cell Res, 2010.

72. Yin, L., et al., *Defective lymphotoxin-beta receptor-induced NF-kappaB transcriptional activity in NIK-deficient mice.* Science, 2001. **291**(5511): p. 2162-5.

73. Dejardin, E., *The alternative NF-kappaB pathway from biochemistry to biology: pitfalls and promises for future drug development.* Biochem Pharmacol, 2006. **72**(9): p. 1161-79.

74. Beinke, S. and S.C. Ley, *Functions of NF-kappaB1 and NF-kappaB2 in immune cell biology.* Biochem J, 2004. **382**(Pt 2): p. 393-409.

75. Sun, S.C., *Controlling the fate of NIK: a central stage in noncanonical NF-kappaB signaling.* Sci Signal, 2010. **3**(123): p. pe18.

76. Varfolomeev, E., et al., *IAP antagonists induce autoubiquitination of c-IAPs, NF-kappaB activation, and TNFalpha-dependent apoptosis.* Cell, 2007. **131**(4): p. 669-81.

77. Vince, J.E., et al., *IAP antagonists target cIAP1 to induce TNFalpha-dependent apoptosis.* Cell, 2007. **131**(4): p. 682-93.

78. Liao, G., et al., *Regulation of the NF-kappaB-inducing kinase by tumor necrosis factor receptor-associated factor 3-induced degradation.* J Biol Chem, 2004. **279**(25): p. 26243-50.

79. Xiao, G., A. Fong, and S.C. Sun, *Induction of p100 processing by NF-kappaB-inducing kinase involves docking IkappaB kinase alpha (IKKalpha) to p100 and IKKalpha-mediated phosphorylation.* J Biol Chem, 2004. **279**(29): p. 30099-105.

80. Razani, B., et al., *Negative feedback in noncanonical NF-kappaB signaling modulates NIK stability through IKKalpha-mediated phosphorylation.* Sci Signal, 2010. **3**(123): p. ra41.

81. Zarnegar, B., et al., *Control of canonical NF-kappaB activation through the NIK-IKK complex pathway.* Proc Natl Acad Sci U S A, 2008. **105**(9): p. 3503-8.

82. Sasaki, C.Y., P. Ghosh, and D.L. Longo, *Recruitment of RelB to the Csf2 promoter enhances RelA-mediated transcription of granulocyte-macrophage colony stimulating factor.* J Biol Chem, 2010.

83. Staudt, L.M., *Oncogenic activation of NF-kappaB.* Cold Spring Harb Perspect Biol, 2010. **2**(6): p. a000109.

84. Brown, K.D., E. Claudio, and U. Siebenlist, *The roles of the classical and alternative nuclear factor-kappaB pathways: potential implications for autoimmunity and rheumatoid arthritis.* Arthritis Res Ther, 2008. **10**(4): p. 212.

REFERENCES

85. Karin, M., *NF-kappaB as a critical link between inflammation and cancer.* Cold Spring Harb Perspect Biol, 2009. **1**(5): p. a000141.

86. Pasparakis, M., *Regulation of tissue homeostasis by NF-kappaB signalling: implications for inflammatory diseases.* Nat Rev Immunol, 2009. **9**(11): p. 778-88.

87. Annunziata, C.M., et al., *Frequent engagement of the classical and alternative NF-kappaB pathways by diverse genetic abnormalities in multiple myeloma.* Cancer Cell, 2007. **12**(2): p. 115-30.

88. Hiscott, J., et al., *Manipulation of the nuclear factor-kappaB pathway and the innate immune response by viruses.* Oncogene, 2006. **25**(51): p. 6844-67.

89. Xiao, G., et al., *Retroviral oncoprotein Tax induces processing of NF-kappaB2/p100 in T cells: evidence for the involvement of IKKalpha.* EMBO J, 2001. **20**(23): p. 6805-15.

90. Atkinson, P.G., et al., *Latent membrane protein 1 of Epstein-Barr virus stimulates processing of NF-kappa B2 p100 to p52.* J Biol Chem, 2003. **278**(51): p. 51134-42.

91. Eliopoulos, A.G., et al., *Epstein-Barr virus-encoded latent infection membrane protein 1 regulates the processing of p100 NF-kappaB2 to p52 via an IKKgamma/NEMO-independent signalling pathway.* Oncogene, 2003. **22**(48): p. 7557-69.

92. Cogswell, P.C., et al., *Selective activation of NF-kappa B subunits in human breast cancer: potential roles for NF-kappa B2/p52 and for Bcl-3.* Oncogene, 2000. **19**(9): p. 1123-31.

93. Weih, F. and J. Caamano, *Regulation of secondary lymphoid organ development by the nuclear factor-kappaB signal transduction pathway.* Immunol Rev, 2003. **195**: p. 91-105.

94. Miyawaki, S., et al., *A new mutation, aly, that induces a generalized lack of lymph nodes accompanied by immunodeficiency in mice.* Eur J Immunol, 1994. **24**(2): p. 429-34.

95. Shinkura, R., et al., *Defects of somatic hypermutation and class switching in alymphoplasia (aly) mutant mice.* Int Immunol, 1996. **8**(7): p. 1067-75.

96. Banks, T.A., et al., *Lymphotoxin-alpha-deficient mice. Effects on secondary lymphoid organ development and humoral immune responsiveness.* J Immunol, 1995. **155**(4): p. 1685-93.

97. Futterer, A., et al., *The lymphotoxin beta receptor controls organogenesis and affinity maturation in peripheral lymphoid tissues.* Immunity, 1998. **9**(1): p. 59-70.

98. Matsushima, A., et al., *Essential role of nuclear factor (NF)-kappaB-inducing kinase and inhibitor of kappaB (IkappaB) kinase alpha in NF-kappaB activation through lymphotoxin beta receptor, but not through tumor necrosis factor receptor I.* J Exp Med, 2001. **193**(5): p. 631-6.

99. Shinkura, R., et al., *Alymphoplasia is caused by a point mutation in the mouse gene encoding Nf-kappa b-inducing kinase.* Nat Genet, 1999. **22**(1): p. 74-7.

100. Ochsenbein, A.F., et al., *Roles of tumour localization, second signals and cross priming in cytotoxic T-cell induction.* Nature, 2001. **411**(6841): p. 1058-64.

101. Lakkis, F.G., et al., *Immunologic 'ignorance' of vascularized organ transplants in the absence of secondary lymphoid tissue.* Nat Med, 2000. **6**(6): p. 686-8.

102. Yamada, T., et al., *Abnormal immune function of hemopoietic cells from alymphoplasia (aly) mice, a natural strain with mutant NF-kappa B-inducing kinase.* J Immunol, 2000. **165**(2): p. 804-12.

103. Jin, W., et al., *Regulation of Th17 cell differentiation and EAE induction by MAP3K NIK.* Blood, 2009. **113**(26): p. 6603-10.

104. Greter, M., J. Hofmann, and B. Becher, *Neo-lymphoid aggregates in the adult liver can initiate potent cell-mediated immunity.* PLoS Biol, 2009. **7**(5): p. e1000109.

REFERENCES

105. Matsumoto, M., et al., *Essential role of NF-kappa B-inducing kinase in T cell activation through the TCR/CD3 pathway.* J Immunol, 2002. **169**(3): p. 1151-8.

106. Ishimaru, N., et al., *Regulation of naive T cell function by the NF-kappaB2 pathway.* Nat Immunol, 2006. **7**(7): p. 763-72.

107. Sanchez-Valdepenas, C., et al., *Nuclear factor-kappa B Inducing Kinase (NIK) is required for graft versus host disease.* Haematologica, 2010.

108. Rennert, P.D., et al., *Essential role of lymph nodes in contact hypersensitivity revealed in lymphotoxin-alpha-deficient mice.* J Exp Med, 2001. **193**(11): p. 1227-38.

109. Karrer, U., et al., *On the key role of secondary lymphoid organs in antiviral immune responses studied in alymphoplastic (aly/aly) and spleenless (Hox11(-)/-) mutant mice.* J Exp Med, 1997. **185**(12): p. 2157-70.

110. de Vos, A.F., et al., *Transfer of central nervous system autoantigens and presentation in secondary lymphoid organs.* J Immunol, 2002. **169**(10): p. 5415-23.

111. Zhang, H., et al., *Intrinsic and induced regulation of the age-associated onset of spontaneous experimental autoimmune encephalomyelitis.* J Immunol, 2008. **181**(7): p. 4638-47.

112. Bettelli, E., et al., *Myelin oligodendrocyte glycoprotein-specific T cell receptor transgenic mice develop spontaneous autoimmune optic neuritis.* J Exp Med, 2003. **197**(9): p. 1073-81.

113. Karrer, U., et al., *Immunodeficiency of alymphoplasia mice (aly/aly) in vivo: structural defect of secondary lymphoid organs and functional B cell defect.* Eur J Immunol, 2000. **30**(10): p. 2799-807.

114. Moyron-Quiroz, J.E., et al., *Persistence and responsiveness of immunologic memory in the absence of secondary lymphoid organs.* Immunity, 2006. **25**(4): p. 643-54.

115. Lund, F.E., et al., *Lymphotoxin-alpha-deficient mice make delayed, but effective, T and B cell responses to influenza.* J Immunol, 2002. **169**(9): p. 5236-43.

116. Fu, Y.X., et al., *Lymphotoxin-alpha-dependent spleen microenvironment supports the generation of memory B cells and is required for their subsequent antigen-induced activation.* J Immunol, 2000. **164**(5): p. 2508-14.

117. Feuerer, M., et al., *Bone marrow as a priming site for T-cell responses to blood-borne antigen.* Nat Med, 2003. **9**(9): p. 1151-7.

118. Moyron-Quiroz, J.E., et al., *Role of inducible bronchus associated lymphoid tissue (iBALT) in respiratory immunity.* Nat Med, 2004. **10**(9): p. 927-34.

119. Dranoff, G., et al., *Vaccination with irradiated tumor cells engineered to secrete murine granulocyte-macrophage colony-stimulating factor stimulates potent, specific, and long-lasting anti-tumor immunity.* Proc Natl Acad Sci U S A, 1993. **90**(8): p. 3539-43.

120. Junt, T., et al., *Antiviral immune responses in the absence of organized lymphoid T cell zones in plt/plt mice.* J Immunol, 2002. **168**(12): p. 6032-40.

121. Gallegos, A.M. and M.J. Bevan, *Central tolerance to tissue-specific antigens mediated by direct and indirect antigen presentation.* J Exp Med, 2004. **200**(8): p. 1039-49.

122. Proietto, A.I., S. van Dommelen, and L. Wu, *The impact of circulating dendritic cells on the development and differentiation of thymocytes.* Immunol Cell Biol, 2009. **87**(1): p. 39-45.

123. Garceau, N., et al., *Lineage-restricted function of nuclear factor kappaB-inducing kinase (NIK) in transducing signals via CD40.* J Exp Med, 2000. **191**(2): p. 381-6.

124. Tamura, C., et al., *Impaired function of dendritic cells in alymphoplasia (aly/aly) mice for expansion of CD25+CD4+ regulatory T cells.* Autoimmunity, 2006. **39**(6): p. 445-53.

125. Lind, E.F., et al., *Dendritic cells require the NF-kappaB2 pathway for cross-presentation of soluble antigens.* J Immunol, 2008. **181**(1): p. 354-63.

126. Sasaki, Y., et al., *NIK overexpression amplifies, whereas ablation of its TRAF3-binding domain replaces BAFF:BAFF-R-mediated survival signals in B cells.* Proc Natl Acad Sci U S A, 2008. **105**(31): p. 10883-8.

127. Wu, L. and K. Shortman, *Heterogeneity of thymic dendritic cells.* Semin Immunol, 2005. **17**(4): p. 304-12.

128. Proietto, A.I., M.H. Lahoud, and L. Wu, *Distinct functional capacities of mouse thymic and splenic dendritic cell populations.* Immunology and Cell Biology, 2008. **86**(8): p. 700-708.

129. Li, J., et al., *Thymus-homing peripheral dendritic cells constitute two of the three major subsets of dendritic cells in the steady-state thymus.* J Exp Med, 2009. **206**(3): p. 607-22.

130. Lu, L.F., et al., *NF kappa B-inducing kinase deficiency results in the development of a subset of regulatory T cells, which shows a hyperproliferative activity upon glucocorticoid-induced TNF receptor family-related gene stimulation.* J Immunol, 2005. **175**(3): p. 1651-7.

131. Marks, B.R., et al., *Thymic self-reactivity selects natural interleukin 17-producing T cells that can regulate peripheral inflammation.* Nat Immunol, 2009. **10**(10): p. 1125-32.

132. Singer, D.B., *Postsplenectomy sepsis.* Perspect Pediatr Pathol, 1973. **1**: p. 285-311.

133. Diamond, L.K., *Splenectomy in childhood and the hazard of overwhelming infection.* Pediatrics, 1969. **43**(5): p. 886-9.

134. Evans, D.I., *Postsplenectomy sepsis 10 years or more after operation.* J Clin Pathol, 1985. **38**(3): p. 309-11.

135. Beltman, J.B., et al., *Lymph node topology dictates T cell migration behavior.* J Exp Med, 2007. **204**(4): p. 771-80.

136. Trickett, A. and Y.L. Kwan, *T cell stimulation and expansion using anti-CD3/CD28 beads.* J Immunol Methods, 2003. **275**(1-2): p. 251-5.

137. Ingvarsson, S., et al., *Antigen-specific activation of B cells in vitro after recruitment of T cell help with superantigen.* Immunotechnology, 1995. **1**(1): p. 29-39.

138. Zinkernagel, R.M., *Immunology taught by viruses.* Science, 1996. **271**(5246): p. 173-8.

139. Senftleben, U., et al., *Activation by IKKalpha of a second, evolutionary conserved, NF-kappa B signaling pathway.* Science, 2001. **293**(5534): p. 1495-9.

140. Suresh, M., et al., *Role of lymphotoxin alpha in T-cell responses during an acute viral infection.* J Virol, 2002. **76**(8): p. 3943-51.

141. Kumaraguru, U., et al., *Lymphotoxin alpha-/- mice develop functionally impaired CD8+ T cell responses and fail to contain virus infection of the central nervous system.* J Immunol, 2001. **166**(2): p. 1066-74.

142. Fagarasan, S., et al., *Alymphoplasia (aly)-type nuclear factor kappaB-inducing kinase (NIK) causes defects in secondary lymphoid tissue chemokine receptor signaling and homing of peritoneal cells to the gut-associated lymphatic tissue system.* J Exp Med, 2000. **191**(9): p. 1477-86.

143. Yanagawa, Y. and K. Onoe, *Distinct regulation of CD40-mediated interleukin-6 and interleukin-12 productions via mitogen-activated protein kinase and nuclear factor kappaB-inducing kinase in mature dendritic cells.* Immunology, 2006. **117**(4): p. 526-35.

144. Zhou, P., et al., *Secondary lymphoid organs are important but not absolutely required for allograft responses.* Am J Transplant, 2003. **3**(3): p. 259-66.

145. De Togni, P., et al., *Abnormal development of peripheral lymphoid organs in mice deficient in lymphotoxin.* Science, 1994. **264**(5159): p. 703-7.

REFERENCES

146. Gajewska, B.U., et al., *Generation of experimental allergic airways inflammation in the absence of draining lymph nodes.* J Clin Invest, 2001. **108**(4): p. 577-83.

147. Lee, B.J., et al., *Lymphotoxin-alpha-deficient mice can clear a productive infection with murine gammaherpesvirus 68 but fail to develop splenomegaly or lymphocytosis.* J Virol, 2000. **74**(6): p. 2786-92.

148. Guo, Z., et al., *Cutting edge: membrane lymphotoxin regulates CD8(+) T cell-mediated intestinal allograft rejection.* J Immunol, 2001. **167**(9): p. 4796-800.

149. Chin, R., et al., *Confounding factors complicate conclusions in aly model.* Nat Med, 2001. **7**(11): p. 1165-6.

150. Kwa, S.F., P. Beverley, and A.L. Smith, *Peyer's patches are required for the induction of rapid Th1 responses in the gut and mesenteric lymph nodes during an enteric infection.* J Immunol, 2006. **176**(12): p. 7533-41.

151. Nakano, H. and M.D. Gunn, *Gene duplications at the chemokine locus on mouse chromosome 4: multiple strain-specific haplotypes and the deletion of secondary lymphoid-organ chemokine and EBI-1 ligand chemokine genes in the plt mutation.* J Immunol, 2001. **166**(1): p. 361-9.

152. Nakano, H., et al., *A novel mutant gene involved in T-lymphocyte-specific homing into peripheral lymphoid organs on mouse chromosome 4.* Blood, 1998. **91**(8): p. 2886-95.

153. Nakano, H., et al., *Genetic defect in T lymphocyte-specific homing into peripheral lymph nodes.* Eur J Immunol, 1997. **27**(1): p. 215-21.

154. Gunn, M.D., et al., *Mice lacking expression of secondary lymphoid organ chemokine have defects in lymphocyte homing and dendritic cell localization.* J Exp Med, 1999. **189**(3): p. 451-60.

155. Mori, S., et al., *Mice lacking expression of the chemokines CCL21-ser and CCL19 (plt mice) demonstrate delayed but enhanced T cell immune responses.* J Exp Med, 2001. **193**(2): p. 207-18.

156. Yasuda, T., et al., *Chemokines CCL19 and CCL21 promote activation-induced cell death of antigen-responding T cells.* Blood, 2007. **109**(2): p. 449-56.

157. Tripp, R.A., et al., *Bone marrow can function as a lymphoid organ during a primary immune response under conditions of disrupted lymphocyte trafficking.* J Immunol, 1997. **158**(8): p. 3716-20.

158. Fu, Y.X., et al., *Lymphotoxin-alpha (LTalpha) supports development of splenic follicular structure that is required for IgG responses.* J Exp Med, 1997. **185**(12): p. 2111-20.

159. William, J., et al., *Evolution of autoantibody responses via somatic hypermutation outside of germinal centers.* Science, 2002. **297**(5589): p. 2066-70.

160. Shintani, S., et al., *Do lampreys have lymphocytes? The Spi evidence.* Proc Natl Acad Sci U S A, 2000. **97**(13): p. 7417-22.

161. Cooper, M.D. and M.N. Alder, *The evolution of adaptive immune systems.* Cell, 2006. **124**(4): p. 815-22.

162. Rogozin, I.B., et al., *Evolution and diversification of lamprey antigen receptors: evidence for involvement of an AID-APOBEC family cytosine deaminase.* Nat Immunol, 2007. **8**(6): p. 647-56.

163. Guo, P., et al., *Dual nature of the adaptive immune system in lampreys.* Nature, 2009. **459**(7248): p. 796-801.

164. Han, B.W., et al., *Antigen recognition by variable lymphocyte receptors.* Science, 2008. **321**(5897): p. 1834-7.

165. Herrin, B.R., et al., *Structure and specificity of lamprey monoclonal antibodies.* Proc Natl Acad Sci U S A, 2008. **105**(6): p. 2040-5.

166. Zapata, A. and C.T. Amemiya, *Phylogeny of lower vertebrates and their immunological structures.* Curr Top Microbiol Immunol, 2000. **248**: p. 67-107.

167. Litman, G.W., J.P. Cannon, and L.J. Dishaw, *Reconstructing immune phylogeny: new perspectives.* Nat Rev Immunol, 2005. **5**(11): p. 866-79.

REFERENCES

168. Pancer, Z. and M.D. Cooper, *The evolution of adaptive immunity.* Annu Rev Immunol, 2006. **24**: p. 497-518.

169. Conticello, S.G., et al., *Evolution of the AID/APOBEC family of polynucleotide (deoxy)cytidine deaminases.* Mol Biol Evol, 2005. **22**(2): p. 367-77.

170. Flajnik, M.F. and L. Du Pasquier, *Evolution of innate and adaptive immunity: can we draw a line?* Trends Immunol, 2004. **25**(12): p. 640-4.

171. Flajnik, M.a.D.P., L., ed. *Evolution of the immune system.* sixth edition ed. Fundamental Immunology by William E. Paul. 2008. 56-124.

172. Rumfelt, L.L., et al., *The development of primary and secondary lymphoid tissues in the nurse shark Ginglymostoma cirratum: B-cell zones precede dendritic cell immigration and T-cell zone formation during ontogeny of the spleen.* Scand J Immunol, 2002. **56**(2): p. 130-48.

173. Marr, S., et al., *Localization and differential expression of activation-induced cytidine deaminase in the amphibian Xenopus upon antigen stimulation and during early development.* J Immunol, 2007. **179**(10): p. 6783-9.

174. Jeurissen, S.H., *Structure and function of the chicken spleen.* Res Immunol, 1991. **142**(4): p. 352-5.

175. Mussmann, R., L. Du Pasquier, and E. Hsu, *Is Xenopus IgX an analog of IgA?* Eur J Immunol, 1996. **26**(12): p. 2823-30.

176. Bernard, D., et al., *Phenotypic and functional similarity of gut intraepithelial and systemic T cells in a teleost fish.* J Immunol, 2006. **176**(7): p. 3942-9.

177. Yasuda, M., et al., *The sheep and cattle Peyer's patch as a site of B-cell development.* Vet Res, 2006. **37**(3): p. 401-15.

178. Berens von Rautenfeld, D. and K.D. Budras, *Topography, ultrastructure and phagocytic capacity of avian lymph nodes.* Cell Tissue Res, 1983. **228**(2): p. 389-403.

179. Akirav M., L.S.a.R.N., *Lymphoid tissues and organs.* Fundamental Immunology, 2008: p. pp 27-55.

180. Lee, S.S., et al., *Hypermutation in shark immunoglobulin light chain genes results in contiguous substitutions.* Immunity, 2002. **16**(4): p. 571-82.

181. Dooley, H., et al., *First molecular and biochemical analysis of in vivo affinity maturation in an ectothermic vertebrate.* Proc Natl Acad Sci U S A, 2006. **103**(6): p. 1846-51.

182. Rumfelt, L.L., et al., *A shark antibody heavy chain encoded by a nonsomatically rearranged VDJ is preferentially expressed in early development and is convergent with mammalian IgG.* Proc Natl Acad Sci U S A, 2001. **98**(4): p. 1775-80.

183. Dooley, H. and M.F. Flajnik, *Shark immunity bites back: affinity maturation and memory response in the nurse shark, Ginglymostoma cirratum.* Eur J Immunol, 2005. **35**(3): p. 936-45.

184. Rijkers, G.T., E.M. Frederix-Wolters, and W.B. van Muiswinkel, *The immune system of cyprinid fish. Kinetics and temperature dependence of antibody-producing cells in carp (Cyprinus carpio).* Immunology, 1980. **41**(1): p. 91-7.

185. Avtalion, R.R., et al., *Influence of environmental temperature on the immune response in fish.* Curr Top Microbiol Immunol, 1973. **61**: p. 1-35.

186. Boudinot, P., et al., *The glycoprotein of a fish rhabdovirus profiles the virus-specific T-cell repertoire in rainbow trout.* J Gen Virol, 2004. **85**(Pt 10): p. 3099-108.

187. Wilson, M., et al., *What limits affinity maturation of antibodies in Xenopus--the rate of somatic mutation or the ability to select mutants?* EMBO J, 1992. **11**(12): p. 4337-47.

REFERENCES

188. Du Pasquier, L., *[The immunity reactions in the tadpole Alytes obstetricans. II. Characterization of immunocytes]*. C R Seances Soc Biol Fil, 1967. **161**(10): p. 1974-7.

189. Du Pasquier, L., et al., *B-cell development in the amphibian Xenopus*. Immunol Rev, 2000. **175**: p. 201-13.

190. Eberl, G., *From induced to programmed lymphoid tissues: the long road to preempt pathogens*. Trends Immunol, 2007. **28**(10): p. 423-8.

191. Grey, H.M., *Phylogeny of the Immune Response. Studies on Some Physical Chemical and Serologic Characteristics of Antibody Produced in the Turtle*. J Immunol, 1963. **91**: p. 819-25.

192. Turchin, A. and E. Hsu, *The generation of antibody diversity in the turtle*. J Immunol, 1996. **156**(10): p. 3797-805.

193. Higgins, D.A., O.K. Ko, and S.W. Chan, *Duck antibody responses to keyhole limpet haemocyanin, human immunoglobulin G and the trinitrophenyl hapten. Evidence of affinity maturation*. Avian Pathol, 2001. **30**(4): p. 381-90.

194. Xu, B., et al., *Lymphocyte homing to bronchus-associated lymphoid tissue (BALT) is mediated by L-selectin/PNAd, alpha4beta1 integrin/VCAM-1, and LFA-1 adhesion pathways*. J Exp Med, 2003. **197**(10): p. 1255-67.

195. Ruddle, N.H., *Lymphoid neo-organogenesis: lymphotoxin's role in inflammation and development*. Immunol Res, 1999. **19**(2-3): p. 119-25.

196. Mooij, P., H.J. de Wit, and H.A. Drexhage, *An excess of dietary iodine accelerates the development of a thyroid-associated lymphoid tissue in autoimmune prone BB rats*. Clin Immunol Immunopathol, 1993. **69**(2): p. 189-98.

197. Ziff, M., *Role of endothelium in chronic inflammation*. Springer Semin Immunopathol, 1989. **11**(2): p. 199-214.

198. Prineas, J.W., *Multiple sclerosis: presence of lymphatic capillaries and lymphoid tissue in the brain and spinal cord*. Science, 1979. **203**(4385): p. 1123-5.

199. Raine, C.S., et al., *Homing to central nervous system vasculature by antigen-specific lymphocytes. II. Lymphocyte/endothelial cell adhesion during the initial stages of autoimmune demyelination*. Lab Invest, 1990. **63**(4): p. 476-89.

200. Serafini, B., et al., *Detection of ectopic B-cell follicles with germinal centers in the meninges of patients with secondary progressive multiple sclerosis*. Brain Pathol, 2004. **14**(2): p. 164-74.

201. Magliozzi, R., et al., *Meningeal B-cell follicles in secondary progressive multiple sclerosis associate with early onset of disease and severe cortical pathology*. Brain, 2007. **130**(Pt 4): p. 1089-104.

202. Campbell, D.A., et al., *Immunohistological analysis of lung tissue from patients with cryptogenic fibrosing alveolitis suggesting local expression of immune hypersensitivity*. Thorax, 1985. **40**(6): p. 405-11.

203. Wallace, W.A., et al., *The immunological architecture of B-lymphocyte aggregates in cryptogenic fibrosing alveolitis*. J Pathol, 1996. **178**(3): p. 323-9.

204. Carlsen, H.S., et al., *B cell attracting chemokine 1 (CXCL13) and its receptor CXCR5 are expressed in normal and aberrant gut associated lymphoid tissue*. Gut, 2002. **51**(3): p. 364-71.

205. Weninger, W., et al., *Naive T cell recruitment to nonlymphoid tissues: a role for endothelium-expressed CC chemokine ligand 21 in autoimmune disease and lymphoid neogenesis*. J Immunol, 2003. **170**(9): p. 4638-48.

206. Kaiserling, E., *Newly-formed lymph nodes in the submucosa in chronic inflammatory bowel disease*. Lymphology, 2001. **34**(1): p. 22-9.

207. Mosnier, J.F., et al., *The intraportal lymphoid nodule and its environment in chronic active hepatitis C: an immunohistochemical study*. Hepatology, 1993. **17**(3): p. 366-71.

REFERENCES

208. Murakami, J., et al., *Functional B-cell response in intrahepatic lymphoid follicles in chronic hepatitis C.* Hepatology, 1999. **30**(1): p. 143-50.

209. Grant, A.J., et al., *MAdCAM-1 expressed in chronic inflammatory liver disease supports mucosal lymphocyte adhesion to hepatic endothelium (MAdCAM-1 in chronic inflammatory liver disease).* Hepatology, 2001. **33**(5): p. 1065-72.

210. Grant, A.J., et al., *Hepatic expression of secondary lymphoid chemokine (CCL21) promotes the development of portal-associated lymphoid tissue in chronic inflammatory liver disease.* Am J Pathol, 2002. **160**(4): p. 1445-55.

211. Klein, I., H.J. Gassel, and I.N. Crispe, *Cytotoxic T-cell response following mouse liver transplantation is independent of the initial site of T-cell priming.* Transplant Proc, 2006. **38**(10): p. 3241-3.

212. Legendre, C., et al., *Transfer of symptomatic peanut allergy to the recipient of a combined liver-and-kidney transplant.* N Engl J Med, 1997. **337**(12): p. 822-4.

213. Tavian, M. and B. Peault, *Embryonic development of the human hematopoietic system.* Int J Dev Biol, 2005. **49**(2-3): p. 243-50.

214. Bradley, L.M., S.R. Watson, and S.L. Swain, *Entry of naive CD4 T cells into peripheral lymph nodes requires L-selectin.* J Exp Med, 1994. **180**(6): p. 2401-6.

215. Lepault, F., et al., *Recirculation, phenotype and functions of lymphocytes in mice treated with monoclonal antibody MEL-14.* Eur J Immunol, 1994. **24**(12): p. 3106-12.

216. Andreakos, E., et al., *Ikappa B kinase 2 but not NF-kappa B-inducing kinase is essential for effective DC antigen presentation in the allogeneic mixed lymphocyte reaction.* Blood, 2003. **101**(3): p. 983-91.

217. Mathis, D. and C. Benoist, *Aire.* Annu Rev Immunol, 2009. **27**: p. 287-312.

218. Dakic, A., et al., *Development of the dendritic cell system during mouse ontogeny.* J Immunol, 2004. **172**(2): p. 1018-27.

219. Brocker, T., M. Riedinger, and K. Karjalainen, *Targeted expression of major histocompatibility complex (MHC) class II molecules demonstrates that dendritic cells can induce negative but not positive selection of thymocytes in vivo.* J Exp Med, 1997. **185**(3): p. 541-50.

220. Bonasio, R., et al., *Clonal deletion of thymocytes by circulating dendritic cells homing to the thymus.* Nat Immunol, 2006. **7**(10): p. 1092-100.

221. Ramsdell, F. and B.J. Fowlkes, *Clonal deletion versus clonal anergy: the role of the thymus in inducing self tolerance.* Science, 1990. **248**(4961): p. 1342-8.

222. Ramsdell, F., T. Lantz, and B.J. Fowlkes, *A nondeletional mechanism of thymic self tolerance.* Science, 1989. **246**(4933): p. 1038-41.

223. Bendelac, A., *Nondeletional pathways for the development of autoreactive thymocytes.* Nat Immunol, 2004. **5**(6): p. 557-8.

224. Birnberg, T., et al., *Lack of conventional dendritic cells is compatible with normal development and T cell homeostasis, but causes myeloid proliferative syndrome.* Immunity, 2008. **29**(6): p. 986-97.

225. Ueno, T., et al., *CCR7 signals are essential for cortex-medulla migration of developing thymocytes.* J Exp Med, 2004. **200**(4): p. 493-505.

226. Proietto, A.I., et al., *Dendritic cells in the thymus contribute to T-regulatory cell induction.* Proc Natl Acad Sci U S A, 2008. **105**(50): p. 19869-74.

227. Zhu, M., et al., *NF-kappaB2 is required for the establishment of central tolerance through an Aire-dependent pathway.* J Clin Invest, 2006. **116**(11): p. 2964-71.

REFERENCES

228. Kajiura, F., et al., *NF-kappa B-inducing kinase establishes self-tolerance in a thymic stroma-dependent manner.* J Immunol, 2004. **172**(4): p. 2067-75.

229. Anderson, M.S., et al., *Projection of an immunological self shadow within the thymus by the aire protein.* Science, 2002. **298**(5597): p. 1395-401.

230. Liston, A., et al., *Aire regulates negative selection of organ-specific T cells.* Nat Immunol, 2003. **4**(4): p. 350-4.

231. Tsubata, R., et al., *Autoimmune disease of exocrine organs in immunodeficient alymphoplasia mice: a spontaneous model for Sjogren's syndrome.* Eur J Immunol, 1996. **26**(11): p. 2742-8.

232. Speirs, K., et al., *Cutting edge: NF-kappa B2 is a negative regulator of dendritic cell function.* J Immunol, 2004. **172**(2): p. 752-6.

233. Xie, P., et al., *Tumor necrosis factor receptor-associated factor 3 is a critical regulator of B cell homeostasis in secondary lymphoid organs.* Immunity, 2007. **27**(2): p. 253-67.

234. Burkly, L., et al., *Expression of relB is required for the development of thymic medulla and dendritic cells.* Nature, 1995. **373**(6514): p. 531-6.

235. Zhang, B., et al., *NF-kappaB2 is required for the control of autoimmunity by regulating the development of medullary thymic epithelial cells.* J Biol Chem, 2006. **281**(50): p. 38617-24.

236. Kobayashi, T., et al., *TRAF6 is a critical factor for dendritic cell maturation and development.* Immunity, 2003. **19**(3): p. 353-63.

237. Zhang, X., et al., *A role for the IkappaB family member Bcl-3 in the control of central immunologic tolerance.* Immunity, 2007. **27**(3): p. 438-52.

238. Akiyama, T., et al., *Dependence of self-tolerance on TRAF6-directed development of thymic stroma.* Science, 2005. **308**(5719): p. 248-51.

239. Drayton, D.L., et al., *Lymphoid organ development: from ontogeny to neogenesis.* Nat Immunol, 2006. **7**(4): p. 344-53.

240. Claudio, E., et al., *BAFF-induced NEMO-independent processing of NF-kappa B2 in maturing B cells.* Nat Immunol, 2002. **3**(10): p. 958-65.

241. Aya, K., et al., *NF-(kappa)B-inducing kinase controls lymphocyte and osteoclast activities in inflammatory arthritis.* J Clin Invest, 2005. **115**(7): p. 1848-54.

242. Tarner, I.H., U. Muller-Ladner, and S. Gay, *Emerging targets of biologic therapies for rheumatoid arthritis.* Nat Clin Pract Rheumatol, 2007. **3**(6): p. 336-45.

243. Becher, B., B.G. Durell, and R.J. Noelle, *Experimental autoimmune encephalitis and inflammation in the absence of interleukin-12.* J Clin Invest, 2002. **110**(4): p. 493-7.

244. Becher, B., B.G. Durell, and R.J. Noelle, *IL-23 produced by CNS-resident cells controls T cell encephalitogenicity during the effector phase of experimental autoimmune encephalomyelitis.* J Clin Invest, 2003. **112**(8): p. 1186-91.

245. Greter, M., et al., *Dendritic cells permit immune invasion of the CNS in an animal model of multiple sclerosis.* Nat Med, 2005. **11**(3): p. 328-34.

246. Becher, B., et al., *The clinical course of experimental autoimmune encephalomyelitis and inflammation is controlled by the expression of CD40 within the central nervous system.* J Exp Med, 2001. **193**(8): p. 967-74.

247. Gutcher, I., et al., *Interleukin 18-independent engagement of interleukin 18 receptor-alpha is required for autoimmune inflammation.* Nat Immunol, 2006. **7**(9): p. 946-53.

248. Wirnsberger, G., F. Mair, and L. Klein, *Regulatory T cell differentiation of thymocytes does not require a dedicated antigen-presenting cell but is under T cell-intrinsic developmental control.* Proc Natl Acad Sci U S A, 2009. **106**(25): p. 10278-83.

References

ACKNOWLEDGEMENTS

I thank Burkhard Becher for giving me the opportunity to accomplish my PhD in his lab. He has always provided a rich atmosphere with a lot of scientific freedom and space to explore my own strengths. I was given responsibilities that allowed me to grow and learn to become an independent scientist, which I am certain will be of great advantage in my future career. Besides his positively challenging demands, he has always aimed to create a friendly, motivating and fun atmosphere in the lab, which made it very enjoyable to work there every day.

I further thank my thesis committee members Manfred Kopf and Adriano Fontana, who were willing to discuss my work in several meetings throughout this PhD period. Both have added different aspects and ideas to the ongoing experimental work, which has been highly appreciated.

A big cheers goes to the entire Becher lab, including all the past and present members. The Becher lab is a very colorful place with different characters from all over the world. I have enjoyed working with all of you and made some long-lasting great friendships in the lab! In particular, I would like to thank Melanie Greter and Florian Mair, who both extensively supported me in these projects.

I compliment my close friends Sabrina, Maya, Sabine and Jana. You have supported me with empathy and shared most moments that made my time in Zürich so incredibly fun!

My deepest appreciation goes to William and my family. Sedulously you listened to my worries and tried to help with supportive words.

i want morebooks!

Buy your books fast and straightforward online - at one of world's fastest growing online book stores! Environmentally sound due to Print-on-Demand technologies.

Buy your books online at
www.get-morebooks.com

Kaufen Sie Ihre Bücher schnell und unkompliziert online – auf einer der am schnellsten wachsenden Buchhandelsplattformen weltweit! Dank Print-On-Demand umwelt- und ressourcenschonend produziert.

Bücher schneller online kaufen
www.morebooks.de

 VDM Verlagsservicegesellschaft mbH
Heinrich-Böcking-Str. 6-8   Telefon: +49 681 3720 174   info@vdm-vsg.de
D - 66121 Saarbrücken   Telefax: +49 681 3720 1749   www.vdm-vsg.de

Printed by Books on Demand GmbH, Norderstedt / Germany